Social Practices and Dynamic Non-Humans

Cecily Maller • Yolande Strengers
Editors

Social Practices and Dynamic Non-Humans

Nature, Materials and Technologies

palgrave
macmillan

Editors
Cecily Maller
Centre for Urban Research
RMIT University
Melbourne, VIC, Australia

Yolande Strengers
Centre for Urban Research
RMIT University
Melbourne, VIC, Australia

ISBN 978-3-030-06381-8 ISBN 978-3-319-92189-1 (eBook)
https://doi.org/10.1007/978-3-319-92189-1

Cover illustration: Pashkov Andrey / Alamy Stock Photo

Printed on acid-free paper

This Palgrave Macmillan imprint is published by the registered company Springer Nature Switzerland AG
The registered company address is: Gewerbestrasse 11, 6330 Cham, Switzerland

Acknowledgements

This book emerged from a symposium titled 'Animals, Automated Devices and Ecosystems: A Symposium on the Agencies of Dynamic Non-humans in Theories of Practice' that we convened in Barcelona, 9–10 October 2015 at RMIT Europe. RMIT University's Centre for Urban Research and RMIT Europe generously supported the event by providing a vibrant venue and delicious catering, both of which were highly conducive to theoretical inspiration and discussion. We invited contributors from a range of disciplines working with dynamic non-human entities or phenomena and social practices to provoke discussion and advance conceptualisations of non-humans in theories of social practice. As many rich and intriguing ideas emerged during the event, developing a book to share these contributions with a wider audience became the logical next step.

We are deeply indebted to generous research fellowships that provided the time and intellectual space to pursue this project and bring it to fruition. Cecily Maller acknowledges her RMIT University Vice Chancellor's Senior Research Fellowship, and Yolande Strengers thanks the Australian Research Council for her Discovery Early Career Researcher Award on the automated smart home.

The contributors to this collection are what made this book possible, and we thank them for their enthusiasm, insights, and timely and patient responses to our original provocations and editorial feedback. We are privileged to have this opportunity to work with such a fantastic collection of scholars as reflected in the quality of the contributions found within these pages. This extends to those scholars who attended the 2015 Barcelona Symposium but were unable to contribute to this book collection (Mike Hazas, Stanley Blue, Yamini Narayanan, Alan Latham, and Giuseppe Salvia).

We acknowledge and thank ourselves for maintaining a remarkably balanced intellectual collaboration and friendship both before and during the completion of this project. The idea for this symposium and book was conceived in a local bar and will probably be celebrated there as well. We are proud to say that we are often in the position of not being able to decide who should be the lead author of our publications, as was the case for the introductory chapter to this book. At a time when academics, and academia, are under increasing pressure, we are grateful to have the support, humour, and intellectual coaching of each other and our fantastic colleagues in the Beyond Behaviour Change Programme and Centre for Urban Research at RMIT University.

Like all book projects of this kind, there is a serious risk of drowning in the sea of forms and publishing procedures involved in navigating the terrain of permissions, formatting, and deadlines. We are very grateful to Sarah Robertson for keeping us buoyant during this time by providing invaluable research and editorial assistance during the book's final stages and successfully project managing our way through the publication process. We also thank the editorial and production teams at Palgrave Macmillan for making this a relatively painless process, especially for meeting our request for the book's cover design.

Lastly but in no diminished way, we acknowledge that in living and working in Melbourne and Victoria, we are in a country belonging to the Wurundjeri people of the Kulin nations. We pay our respects to elders of the past, present, and future and recognise that their lands and waters were never ceded.

Contents

Notes on Contributors

Paula Arcari *(PhD Candidate/Research Assistant, Beyond Behaviour Change Research Programme, Centre for Urban Research, RMIT University, Melbourne, Australia)* has a background in climate and environmental science. Her research uses the emergence of so-called ethical and sustainable meat to explore how meat consumption and the use of animals as food continue to be (re)constituted as enduring and resilient parts of social practices.

Jennifer Atchison *(Senior Lecturer, Australian Centre for Cultural Environmental Research, School of Geography and Sustainable Communities, University of Wollongong, Australia)* focuses research on the material interactions of human-nature relationships. She is primarily interested in the ways non-humans (particularly plants, but also others) shape human lives and provoke us to think differently and live more sustainably in the world.

Kirsten Gram-Hanssen *(SBI, Danish Building Research Institute, Aalborg University, Copenhagen, Denmark)* leads the research group on Sustainable Housing and Cities in the Danish Building Research Institute, and her main research interest for the past 15 years is on sustainable housing, everyday life practices, and energy consumption.

Sabrina Hauser *(PhD Candidate/Research Assistant, School of Interactive Arts and Technology, Simon Fraser University, Canada, and Instructor in Interaction Design, Emily Carr University of Art and Design, Canada)* is an interaction design researcher with a background in design and information science. Her PhD research explores the intersections between postphenomenology and interaction design research.

Gay Hawkins *(Research Professor in Social and Cultural Theory, Institute for Culture and Society, Western Sydney University, Sydney, Australia)* researches in the areas of materials and more-than-human politics, environmental humanities, and governing everyday life. She brings to this research a background in political theory, cultural studies, and STS. She has done major studies on waste and everyday ethics; the rise of bottled water markets and new drinking practices; and plastics, carbon economies, and issue activism.

Lenneke Kuijer *(Assistant Professor, Department of Industrial Design, Eindhoven University of Technology, Netherlands)* is an Assistant Professor in the Department of Industrial Design at Eindhoven University of Technology. Her expertise lies at the touching points of design and social science, in particular in social practice theory and research through design.

Cecily Maller *(Vice Chancellor's Senior Research Fellow, Beyond Behaviour Change Research Programme, Centre for Urban Research, RMIT University, Melbourne, Australia)* is a co-leader of the Beyond Behaviour Change Research Programme at the Centre for Urban Research, RMIT University, where she researches human-environment interactions, sustainability, and health using post-humanist theories, including theories of social practice.

Janine Morley *(Senior Research Associate, DEMAND Centre (Dynamics of Energy, Mobility and Demand), Department of Sociology, Lancaster University, UK)* is a sociologist who investigates the roles of technology in social change, social practices, and sustainability. Her current research focuses on the changing nature of digital devices, services, and infrastructures and the implications for energy demand.

Doenja Oogjes *(PhD Student/Research Assistant, School of Interactive Arts and Technology, Simon Fraser University, Canada)* is an interaction design researcher with a background in industrial design. Her research explores the design of digital domestic technologies using speculative design and integrating indirect ways technologies mediate our everyday.

Elspeth Oppermann *(Research Fellow, Northern Institute, Charles Darwin University, Darwin, Australia)* is a human geographer, specialising in adaptation to extreme weather and climate change. Her current focus is on the production and management of heat stress in the practices of the labour-intensive, outdoor workforce in hot and humid environments.

Jenny Rinkinen *(Senior Research Associate, DEMAND Centre (Dynamics of Energy, Mobility and Demand), Department of Sociology, Lancaster University, UK)* is a Senior Research Associate in the DEMAND Research Centre at Lancaster University. Her research focuses on material and temporal dimensions of social practices in understanding the dynamics of energy demand.

Yolande Strengers *(Senior Research Fellow, Beyond Behaviour Change Research Programme, Centre for Urban Research, RMIT University, Melbourne, Australia)* is a co-leader of the Beyond Behaviour Change Research Programme at the Centre for Urban Research, RMIT University. Her research is interested in how smart technologies are changing everyday life and energy demand.

Ron Wakkary *(School of Interactive Arts and Technology, Simon Fraser University, Canada, and Professor in Industrial Design, Eindhoven University of Technology, Netherlands)* is an interaction design researcher, and his work investigates the changing nature of interaction design in response to everyday design and social practices.

Gordon Walker *(Co-director of the DEMAND Centre (Dynamics of Energy, Mobility and Demand) and Chair in the Lancaster Environment Centre, Lancaster University, UK)* has expertise on the social and spatial dimensions of sustainable energy technologies, sustainability transitions, sustainable social practices and cross-cutting issues, and theories of energy and environmental justice. He has led a series of multi-partner projects funded by UK research councils and government departments.

List of Figures

List of Tables

1

Dynamic Non-humans in a Changing World

Yolande Strengers and Cecily Maller

The robots are coming! So too is the 'age of automation', the 'internet of things', the 'fourth industrial revolution', the march of 'invasive' species across the planet, more intense natural disasters and extreme temperatures, and a potential cataclysm of other unprecedented events and phenomena of which we do not yet know, and cannot predict. Possible utopian and dystopian scenarios are not solely the result of human activity but involve numerous 'non-humans' that create ripples in, and can also dramatically and irrevocably change, the world as we know it. Despite its limitations in capturing 'the multiplicitous ways of (non)life' (Lulka 2009, 383), the term 'non-humans' refers to the vast array of actants and agents that are enrolled in and affect change across different registers of what humans understand as society, or 'the social'.[1] Social theorists and scholars have long been fascinated by non-human agency, but there is

[1] See Stengers (2010) for a more detailed critique of 'non-human'.

Y. Strengers (✉) • C. Maller
Centre for Urban Research, RMIT University, Melbourne, VIC, Australia
e-mail: yolande.strengers@rmit.edu.au; cecily.maller@rmit.edu.au

C. Maller, Y. Strengers (eds.), *Social Practices and Dynamic Non-Humans*,
https://doi.org/10.1007/978-3-319-92189-1_1

renewed interest in these curious characters and our relationships with them (e.g., Braun and Whatmore 2010a; Haraway 2016; Gandy 2003). A host of questions are being asked, such as: What exactly are non-humans in relation to humans? How can we best 'know', classify, and understand various types of non-human roles and agencies? How are non-humans changing the world, such that the boundaries between what is and isn't human are increasingly blurred? And what do, can, and should we humans do about non-humans, if anything?

Our focus on *dynamic* non-humans invites enquiry about devices, objects, phenomena, and bodies that appear to have capacity to undertake seemingly independent actions or operate autonomously. Examples include automated devices and software, heatwaves, robots, animals, and plants. They are 'dynamic' in the sense that they are constantly changing, progressing, or 'acting up', often in ways that humans have minimal control over, understanding of, or input into. We apply the term loosely, and necessarily imprecisely, with the understanding that all non-humans might be categorised as 'dynamic' in one way or another, and that their dynamism likely varies according to different practices, and temporal and spatial scales. This apparent dynamism inspired this collection, raising questions regarding the apparently increasing autonomy, pervasiveness, and potential for extreme consequences that some non-humans pose. We argue that dynamism warrants special attention, or re-attention, through a detailed understanding of non-humans' positionality, relationality, and roles within social life.

Such investigations are not altogether new. Despite a long history in some disciplines such as science and technology studies (STS), many fields of social enquiry are now beginning to question the absence, or limited presence, of a variety of non-humans from the conceptualisation, interrogation, and analysis of social worlds and in understanding complex contemporary problems like climate change. Through shifts such as the (re)turn to materials (Whatmore 2006), or 'new materialisms' (Bennett 2010; Coole 2013), the 'missing masses' once lamented by Latour (1992) are increasingly making their presence felt in disciplines such as political ecology (e.g., Bennett 2010), human and cultural geography (e.g., Anderson and McFarlane 2011), science and technology studies (e.g., Latour 2005; Haraway 1997), sociology (e.g., Meloni 2014), and human-computer interaction design (e.g., Pierce and Paulos 2010). This burgeoning interest is partly a corrective move to decentre humans as the only permissible social actors (Pickering 2008) and partly due to the growing realisation that global challenges such as antibiotic resistance, widespread digital surveillance, and terrorism are not caused by humans alone, but are the product of entangled relations with myriad other, non-human, actors (Greenhough 2010; Haraway 2016; Maller 2018). These dynamic non-humans have the potential to, and do, create chaos despite our best efforts to control or respond to them.

Recognising that dynamic entanglements of humans and non-humans are ubiquitous in social worlds has given rise to 'more-than-human' thinking and theories in the social sciences and humanities (e.g., Abram 1997; Bell et al. in press, Whatmore 2002). The term 'more-than-human' seeks to move beyond only considering humans and human agency to acknowledge the many human and non-human relations at play in contemporary challenges and everyday life. In this way, more-than-human theories are 'relational ontologies' used to understand the connections between humans and the many non-humans on which we depend and with whom we co-exist (Castree 2003). For example, this includes the millions of bacteria and microorganisms that live in, and on, our bodies and homes and without which we would not survive (Yong 2016). More-than-human approaches often draw together knowledge from the social and natural sciences to understand in some detail how non-humans make and shape the world (Maller 2018). Reflecting on these developments towards inter- and cross-disciplinarity, Head (2016) observes the 'natural' sciences are turning their attention towards people, while the social sciences are more thoroughly and consistently exploring 'nature'. Because crossing disciplinary boundaries is relatively new, 'scholars are still groping with the implications of these findings and new perspectives … [where] the previously neat categories of technology, or materials, are giving way to more lively configurations of matter' (Head 2016, 70). The rise of more-than-human thinking is a clear signal that the dynamism of many existing and new non-humans requires, and is generating, renewed attention.

Following the footsteps (or pawprints) of this messy but exciting scholarly territory, this collection delves into one particular theoretical domain where the dynamism of non-humans has attracted some, but so far, limited attention. Theories of social practice have been variously seen to originate in the work of thinkers such as Wittgenstein, Heidegger (Rouse 2007; Schatzki 2016), and Foucault (Rouse 2007; Schatzki 2001). Practice theories were further developed by Giddens (1984), Bourdieu (1990), and Ortner (1984), who each homed in on the practical nature of action, arguing that individual agency and social structure were co-constitutive, and produced through practices. In later theoretical articulations, scholars understand the social world as comprised of recognisable 'practices' or practice 'entities', activities that are regularly carried out by multiple human actors over time and space (Schatzki 2001). Practices include many routine activities, such as commuting, cooking and eating, studying, gardening, socialising, as well as professional and organisational practices.

Since the late 1990s and early 2000s, the materially focused, post-humanist[2] ideas of Shove and colleagues (Shove 2003; Shove et al. 2012), Schatzki (2002, 2010), Reckwitz (2002a, b), Pickering (1995), and Warde (2005) have been increasingly influential in practice scholarship about contemporary, complex, social and environmental problems. Following developments in STS and actor-network theory (ANT) (e.g., Latour 1992; Law and Hassard 1999), researchers working with these ideas have turned their attention to the material aspects of practices, and the essential roles that non-humans such as appliances, technologies, infrastructures, and devices play in everyday activities from heating or cooling homes and workplaces (Hitchings 2013; Strengers and Maller 2011), washing bodies (Browne et al. 2013), playing sport and doing hobbies (Shove and Pantzar 2005, 2007) and other forms of everyday consumption (Gram-Hanssen 2009; Røpke 2009), through to socio-technical transitions or 'systems of practice' (Watson 2012; Walker and Shove 2007).

Despite connecting with post-humanist and more-than-human agendas, recent scholarship grounded in theories of social practice has only gone so far in its treatment of, and interest in, non-humans. Relatively static,

[2] The post-humanist stance in theories of practice we are referring to is explained by Schatzki (2001). It relates to a focus on practices as the subject of enquiry rather than individual humans and also acknowledges that a variety of non-humans contribute to and constitute human sociality.

bounded, and easily identifiable material objects and technologies, such as showers, air conditioners, freezers, and buildings, have been the dominant focus rather than ecological systems, heat, air, animals, plants, and artificial intelligence.[3] However, as noted by others, living, life-like, or otherwise *dynamic* non-humans carry, express, and share roles and agencies in sometimes confounding and contradictory ways (Bennett 2010; Coole 2013; Strengers et al. 2016) and matter for how practices are performed, emerge, endure, and decline.

One of the starting points for this collection is therefore that the 'human-like' characteristics and agencies of non-humans, like companion animals (Strengers et al. 2016) and robots (Petersen et al. 2017), challenge traditional classifications of 'materials', 'objects', or 'material arrangements' in theories of social practice. Indeed, despite growing recognition of non-human agencies and their capacity to create change and intervene in social worlds, in the broad body of work referring to social practices, the 'carriers' of practices remain decidedly human (e.g., Shove et al. 2012; Simonsen 2012). This becomes potentially problematic when non-humans become 'dynamic' (or perhaps more dynamic than usual), arguably acting performatively, or generally exerting agencies and occupying roles that cannot be confined to common designations, such as an element or organising arrangement of practices. In some ways, maintaining that only humans can carry or perform practices (albeit with material participation) is understandable, if not sensible. For example, in seeking to understand practices of energy use in the context of environmental justice, it makes sense to explore the inequalities and poverty embodied in the social practices as carried by people, within which 'things' and resources participate. However, as momentum builds towards the notion that social worlds can no longer bifurcate people from things, nature from culture, and living from dead matter (Houston et al. 2017; Maller 2018), such a stance becomes complicated.

The complication extends to the kinds of challenging political questions dynamic non-humans give rise to, including their capacity to

[3] This is despite some theories of practice accounting for non-humans under the banner of 'nature', for example, Schatzki (2010, 129) describes his concept of 'material arrangements' as having four interconnected, material entities: 'humans, artifacts, organisms, and things of nature'.

intervene in social worlds—and the events they catalyse (Braun and Whatmore 2010b). For example, can responsibility be assigned to electricity networks and generators during large-scale blackouts where tens of thousands of homes are without power?[4] Are the processes in nature that initiate flooding or wildfire events 'culpable' in the devastating damage caused to human lives, livelihoods, and settlements? Should urban non-human species such as flying foxes (a large, fruit-eating bat) have roosting rights in cities, despite their presence in large numbers destroying highly valued, heritage-listed vegetation? Can dynamic non-humans like urban rivers be assigned rights when their existence and vitality is threatened by human activities? To assist in understanding these and other questions, we need theoretical resources that help make sense of the complex and unpredictable characteristics of human-non-human, or more-than-human, relations. As Braun and Whatmore (2010b, x) highlight in their discussion of 'the force of things', without such conceptual resources we are too often left 'oscillating between fearful repudiation and glib celebration' when it comes to understanding collectives of human-non-human relations.

Responding to this broader challenge, this book seeks to explore both the dynamism of non-humans and the residual human-centrism of theories of practice, issues that also still trouble disciplines such as STS, post-humanist sociology, and human geography. In inviting interdisciplinary contributions to this collection, we posed two main questions for scholars to consider. First, how can 'dynamic' non-humans be conceptualised in theories of social practice? And second, can non-humans be conceived as carriers or performers of practices in their own right? These questions were designed to challenge and generate playful and provocative thought experiments that push the conceptual boundaries and limitations of practice theory scholarship in the authors' own disciplines or subdisciplines. In orienting their gaze towards dynamic non-humans in theories of social practice, contributors address and contest these questions, as well as other theoretical sticking points, to reveal new and insightful directions for understanding, analysing, and potentially intervening in practices for social and environmental change.

[4] For a detailed discussion of this point, see Bennett (2010).

As chapter authors point out, some of the ideas in this book may cause controversy and further debate. For example, whether animals or technologies can carry out practices may be considered a moot point by some scholars, as their activities may simply be rendered 'asocial', that is, not part of the realm of the social and therefore not of relevance in the ways being proposed. However, after Taylor (1993), our stance is that such provocations could enhance our understandings of social practices in relation to the sorts of questions and contemporary issues to which they are applied, including climate change, invasive species management, and the increasing prevalence of autonomous technologies. As Taylor (1993, 464) writes, 'a social science which can assist in the understanding of practices conventionally labelled asocial … is a discipline which has undeniably expanded its explanatory powers'. Social practice theories already go a long way to doing just that, but here we ask if they can go further.

The book is structured by two interrelated parts, each dealing with specific questions related to different relations and interactions of dynamic non-humans in social practices.

Part I is titled 'Nature, Materiality and Processes'. This first part turns attention towards 'natural' non-humans in theories of social practice. Its focus is on practices that involve living, organic, or life-like non-humans, and the biological processes that they are implicated in, and upon which many practices depend. Included are chapters on plants, animals, and other non-humans with a biological, vital, or vibrant material basis, including water and thermodynamic flows of energy. Contributors focus on practices that involve (potentially multiple) bodies and materialities of both humans and dynamic non-humans—for example, those involved in invasive plant management, timber harvesting and wood heating, eating, drinking, working, and sleeping. In doing so, Part I probes the potential boundaries, crossing points and overlaps between biological and social worlds. In turning to questions of biology, Part I also considers what is, and isn't, a practice. For example, are bodily activities or doings such as breathing or sleeping able to be classified as practices, or merely biological processes that can unproblematically remain in the background of social analyses? Other questions considered by contributors include: How can 'dynamic' living or organic non-humans and their capacities, vitalities, and agencies be conceptualised in theories of social practice to redress human-centrism, and move beyond simple classifications of materials, elements, or objects? Is there room for greater blurring of biological and social, human and non-human distinctions in practice-related work, for example, by learning from other theoretical resources such as Haraway's nature-cultures (2008), Whatmore's hybrid geographies (2002), Barad's (2003) concept of intra-action, and Ingold and Palsson's biosocial becomings (2013)? And what are the potential benefits and challenges of delving into these intersections in practice scholarship?

Exploring biosocial becomings and human-plant relations in the context of the mass invasion of 'weeds' during the Anthropocene, Jennifer Atchison (Chapter 2) begins by questioning the assumed passivity of plants through the notion of 'planty practices'. Drawing on theories of practice, new materialisms, and the work of Barad and Pickering, this chapter questions the capacity of some varieties of social practice to adequately account for the non-human agency of living beings, and the notion that only people perform practices. To understand the 'contours of plant agency', Atchison uses empirical examples drawn from ethnographic work on invasive plants in Northern Australia to argue that plants routinely contest and resist practices of natural resource management and control. Atchison shows how invasive plants respond to, and reshape, the practices of invasive plant managers and management processes, rather than simply having practices 'done to' them by human practitioners. Instead of managing the 'risks' that mass 'invasion' seemingly presents and the use of killing and eradication measures as normalised responses, turning to practices provides an opportunity to understand how human-plant relations are producing new socio-material worlds and highlights the realities and complexities of living with invasive life. Atchison concludes that plants demonstrate multiple agencies, including defying attempts at containment or bounding, and form relations with people and other non-humans that can transform ecosystems with global consequences.

Following similar thought experiments to Atchison, Jenny Rinkinen (Chapter 3) draws inspiration from ecological anthropologists such as Ingold, to invoke greater sensitivity towards living organisms and their treatment in social practices. She highlights the current emphasis on 'human-made' materiality over biological and 'natural' materials and processes embedded in non-humans like trees and wood. Living, or formerly living, non-humans such as trees and wood products are often considered as resources in social practices, which potentially mutes and overlooks their vitality and liveliness. Rinkinen explores these ideas through an empirical analysis of small-scale wood heating and harvesting practices in Finland. Because trees have rhythms and are subject to biological processes, in which practices of wood management and household heating practices are embedded, she makes the case that an overly functional view of living organisms overlooks the uncertainty, variation, and sometimes contested roles some non-humans have in practices. For example, as trees are part of ecosystems before they are transformed into wood, they reflect and are characterised by local ecological conditions. This means that wood heating practices carried out at different sites may vary according to the local trees and wood harvesting practices. Arguing that biological and social activity are intertwined in constant processes of becoming, Rinkinen shows how analysing biological processes upon which some practices depend may assist in understanding how practice varieties develop at some sites and locations, but are more standardised across others. Turning to biological and ecosystem processes can therefore help illustrate how practices are locally anchored and vary according to their materialities.

In the next chapter, Paula Arcari (Chapter 4) continues the theme of how best to include living beings in theories of practice. In particular, she explores to what extent practice theories can address deep and sustained inequities and power imbalances between humans and sentient practice 'materials' such as non-human animals. Arcari argues that due to inherent power relations between human and non-human animals that are reproduced by practices, there is a need to more accurately, and ethically, represent living non-human 'others' in this body of theory. She reasons that positioning non-human animals in social practices as 'dynamic' is insufficient as it elides ongoing and problematic relations between humans and non-human animals. She centres her critique on the human-centrism of social practice theories while interrogating the suggestion that non-humans can be performers of practices. Unlike other authors in the collection (Chapters 2, 3, 5, 10, 11), Arcari does not support the notion that non-human animals might perform or co-perform practices. Rather than attempting to broaden or 'enlighten' social practice theories by conceiving non-humans as practice performers, Arcari has an alternative offering. She proposes the concept of the subaltern as a productive way to make visible and give 'voice' to non-human animals, directly tackling human-centrism and the at times ethically neutral stance of social practice theories.

Cecily Maller (Chapter 5) shifts the discussion from animal to human bodies by examining how to represent our own bodily materiality in theories of practice. She focuses on two issues relevant to investigating topics such as health and wellbeing, and concerns about how both human and non-human bodies can be impacted by social practices. Her first contention is that many aspects of bodies that make them dynamic can be under-acknowledged in practice scholarship. Second, she argues that social practice theories are concerned with a narrow definition of bodies, where only human bodies seemingly count. Maller draws on scholarship from political ecology, new materialisms, and assemblage thinking to introduce two ways human and non-human bodies could be reconceptualised in theories of practice. These are bodies as vibrant materials in practices, and bodies as more-than-human assemblages. In doing so, she picks up Shove's suggestion that there is scope to distinguish the qualities and characteristics of different materialities, according to how directly or indirectly they are implicated in practices. In starting to unpack what bodies do in practices, and what practices do to bodies, Maller suggests there may be ways to extend the reach of theories of practice, enabling them to turn to political questions related to equity, justice, and health. She concludes that reframing bodies in social practices, and paying more attention to their biosocial dimensions, could be valuable in understanding how practice trajectories might shift what this might mean for social and environmental challenges.

In her chapter, Gay Hawkins (Chapter 6) eloquently illustrates the agency of bodily-material relations and their capacity for global practice change, invoking new products, materialities, and accountabilities, which in turn create new ethical and political dilemmas. Bringing in thinking from STS and the work of Mol, Woolgar, and Neyland, she analyses how the mobile practice of carrying and sipping water from branded plastic bottles has spread around the world. With an interest in ontological politics, Hawkins asks how objects 'come into being' through practices and how these practices provoke new realities and politics. She discusses how accountability is produced by, and can be assigned to, materials and their various relations between people and non-humans. In doing so, Hawkins shows how practices can produce ontologies through reconfiguring both materials and bodies, and the wider systems and complexes of practices of which they are a part. In reviewing the rise of mobile drinking and its socio-material relations, she considers how this practice diverges, converges, or interferes with existing ways water is delivered and consumed. Hawkins then questions the implications of commoditised mobile drinking, and its associated markets and beverage corporations, for more sustainable, democratised drinking water supply delivered by networked pipes and taps. She concludes by arguing that the marketing of plastic water bottles and the branding of water products has reconfigured material relations between the practice of drinking and thirsty bodies, where a practice that was largely performed 'on the move' is now common in homes and workplaces, displacing unbounded, unmarketed tap water with a branded, bounded, and wasteful commodity.

Elspeth Oppermann and Gordon Walker (Chapter 7) focus their attention on thermodynamic flows and thermal energies that, they argue, all practices depend on and intervene in. In their interest in energetic-material relations, in particular heat, they introduce a different type of dynamic non-human into the book's discussion so far. Drawing on Schatzki's conceptualisations of material arrangements and flows, Oppermann and Walker posit that rather than being categorised as a material element of practice, heat is instead a dynamic energy in which all practices are immersed and that heat is both inherently productive of, and produced by, practices. Thermodynamic flows, or energetic-material relations, occur through a range of processes, including through practices of hastening, retarding, redirecting, collecting, converting, or producing thermal energy, all of which can affect the comfort and optimal performance of humans, non-human animals, plants, and machines. Oppermann and Walker contend that excesses and deficits of heat, as in when bodies feel hot or cold or machines function well or not, foreground these flows that can be detected by the senses and/or produce material affects. Achievement of homeostasis, or a comfortable thermal operating range, they argue, is therefore a constant practice. Oppermann and Walker work through their conceptualisation of heat, its materialities, and impacts regarding the social practices of outdoor manual workers in Northern Australia's tropical monsoon zone. They find that in outdoor, labour-intensive work in extreme, humid conditions, the heat that workers absorb, generate, and lose through rhythms of work practices is co-constitutive of, and constituted by, temporal patterns of bodily heat storage and loss. Oppermann and Walker conclude by calling for material-energetic relations to be conceptualised more expansively within applications of social practice theory, thereby rendering the invisible visible and enabling more effective management of thermal energies and thermodynamic flows in everyday life.

Part II is 'Technologies, Automation and Performativity'. This part of the book moves away from the usual suspects of appliances, objects, and other easily bounded things and turns attention towards some less common non-living materials in theories of social practice. It asks whether there is something distinctive about technologically advanced dynamic non-humans—such as robots, automated devices, and software—which allow them to perform, carry out, or otherwise participate in the enactment of social practice in ways which extend beyond conventional understandings of their role. As such, this section of the book necessarily engages with questions about the very sociality of social practices. Contributors challenge, defend, or deflect the idea that a human performer or bodily carrier must be present for and central to the enactment of social practice. This part of the book is therefore concerned with boundaries between who, or what, can or cannot carry out social practices and how to account for the performative roles of dynamic non-humans within post-humanist theories of social practice.

All contributors to Part II engage with dynamic non-humans which in some way appear to step outside the 'normal' roles usually reserved for materials in theories of practice. Robots and automation devices, for example, are often explicitly positioned (by humans) as taking over from or carrying out human activity. However, this is not merely a case of delegation. Autonomous technologies and infrastructures may take on 'a life of their own', acting back or performing practices in unexpected ways and potentially carrying their own 'doings and sayings'. Key questions for these contributors are whether there is anything distinctive about some non-humans which warrant them status as a performer or carrier of social practice? Do existing accounts of materiality in theories of social practice adequately account for the disruptive and dynamic agencies of these 'human-like' non-humans? And what qualities make some dynamic non-humans distinctive from those assigned the role of elements, arrangements, or some other material process or relationship to practice?

Ron Wakkary, Sabrina Hauser, and Doenja Oogjes (Chapter 8) approach these questions by exploring the role of relatively banal and seemingly inconsequential things in everyday domestic practices drawing on their design research and concepts from the philosophy of technology. Through some experiments with invented research artefacts, they develop the concept of displacement, which captures the 'unknowability' of materials and non-humans that are used, and often taken for granted, in everyday practices. In doing so they challenge the idea that non-humans have to be of a spectacular, advanced, or otherwise distinctive nature to warrant further attention to their dynamism. Instead, they use their research to set up a series of obscure relations between things that are, or can be, outside human perception or lacking a clear role or purpose in any identifiable practice. Ceramic bowls and cups that communicate to each other over the internet in Morse code, and a 'table-non-table' which is a moving stack of paper on top of an aluminium chassis, are at once delightfully pointless and conceptually intriguing things which are difficult to 'place' in theories of practice. It is this 'displacement' that interests Wakkary and colleagues and allows for their introduction of the post-phenomenological concepts of withdrawal and multistability to understand these 'material speculations' in design research. They argue that displacement shifts current ideas of practice to bring things, or material elements, into greater relief and contrast. By 'backgrounding' practices, these authors posit that detailed fragments of the performances of practices are revealed that would otherwise go unnoticed. Wakkary, Hauser, and Oogjes conclude that even when non-humans are outside of human perception, they still have dynamic effects on and with practices that are deserving of our attention.

Janine Morley (Chapter 9) turns our attention to software and automation, which present as 'challenging' materials to conceptualise and study in theories of social practice. Both have the ability to decentre the role of human bodily activity and exhibit particularly unpredictable, responsive, and dynamic qualities on a large scale. Despite their unique traits, Morley mounts a convincing case that these and other dynamic non-humans can, and should, be accounted for within existing theories of social practice, albeit with their own potentially distinctive role. Drawing on the examples of self-driving and connected cars, Morley picks up Shove, Pantzar, and Watson's concept of 'connective tissue' to zoom outwards from 'distracting' questions about how software is embedded within individual practice performances, focusing instead on larger instantiations of how software holds together (and pulls apart) social practice complexes. Through this analysis Morley draws attention to the modes of 'feedback' that extend beyond individual users or the various stakeholders involved in producing self-driving and connected cars, to the connections between practices that software binds together. She shows how software contributes to making and re-making relationships between practices by extending what human bodies can do within existing social practices. Morley's contribution provides fruitful insights for analysing 'large' contemporary socio-technical concerns, such as the changing system of automobile practices, and how they reproduce, change, and extend across time with the introduction of increasingly dynamic non-humans.

Homing in on the increasing automation of everyday household artefacts and appliances, Lenneke Kuijer (Chapter 10) puts forward the concept of co-performance to account for the practices people perform *with* dynamic non-humans. Kuijer situates her 'thought experiment' in STS and human-computer interaction design, focusing on how laundering has changed over time from a manual to an automated and ultimately co-performed practice between humans and machines. She argues that automated artefacts are capable of some forms of performance, or 'artificial doings and sayings', due to the increasing role of sensors, actuators, and computing power allowing devices to make decisions and act autonomously in some circumstances, including in ways that exist outside of direct human perception. However, because this 'artificial performance' is only meaningful as part of social practices, Kuijer stops short of suggesting that non-humans can perform practices on their own. Her concept of co-performance brings a fresh perspective to our historical understanding of how laundering has changed over time, as a practice 'career' which has involved the accumulated learning not only of human body-minds but of 'artificial body-minds'. She concludes by suggesting that the concept of co-performance could be helpful for designers in understanding how automated devices contribute to social practice change.

Yolande Strengers (Chapter 11) explores the performative potential of service robots in theories of social practice, which are expected to 'take over' increasingly sophisticated tasks previously performed by humans. She asks whether this intended and possible reality requires rethinking the role of some dynamic non-humans as being capable of performing social practices. Strengers explores this question by tracing the promises for, and early household experiences with, robotic vacuum cleaners, or robovacs. In doing so, she comes across some other dynamic non-humans—namely, household pets—that appear to be riding robovacs for fun. She speculates on three possible roles for these service robots in theories of practice: first, as a 'standard' material element or device in vacuuming practices still carried out by humans (albeit with changing scales, responsibilities, and roles); second, as an (actual or perceived) non-human performer of robo-vacuuming; and third, as an element of another non-human practice known to humans as 'Roomba Riding', in which animals ride robovacs for their own entertainment. Rather than 'picking' one perspective, Strengers follows Rinkinen and Shove in calling for a relational and provisional understanding of non-humans in theories of social practice, depending on the empirical and conceptual questions at hand. She argues that each provides valuable and different understandings of our changing relationship with artificial intelligence such as service robots.

Kirsten Gram-Hanssen (Chapter 12) draws on the symmetrical anthropology of Latour alongside Schatzki's theory of practice to propose that both humans and some dynamic non-humans (like automation) can perform practices. She justifies this move by drawing attention to the increasing prevalence and dynamism of automation in everyday life, approaching her analysis through the example of the smart home. More specifically, she contends that Latour's attribution of equal agency between humans and non-humans can be extended to other practice theory concepts which she develops in her chapter. Gram-Hanssen proposes three possible ways to approach this suggestion. First, she argues that Schatzki's concept of 'dispersed practices' can be employed to understand the distinctive practices that 'simple' automated non-humans perform (such as turning something on and off), which in turn are embedded within 'integrative practices' that humans carry and perform (such as heating a room). Second, she explores the possibility that automated technologies with self-learning algorithms can also perform their own integrative practices and pursue their own ends and rules. Finally, she explores the idea that larger-scale automated systems, like 'smart' energy systems, can be thought of as part of the 'material infrastructure' of energy-consuming practices *and* performers of practices in their own right. Each perspective provides a unique approach to analysing the introduction, intensification, and increasing independence of automation into everyday life.

Conclusion

Myriad dynamic non-humans are implicated in current and future changes, and challenges, to everyday life across the globe. From profound developments in artificial intelligence, nanotechnology, and automation to revelations about the interdependence of humans with all other life forms, it appears that the lines between humans and non-humans are being increasingly blurred and contested. With their post-humanism and detailed understandings of the intricacies of everyday life, theories of social practice are well-positioned to frame and answer questions about how dynamic non-humans are changing the world and ourselves. Yet to date this potential has only been partially realised. This is partly the result of a dominant focus in theories of practice on understanding, unpacking, and reframing problems of resource use and consumption by exploring consumption practices performed by people. However, as has happened in cognate theoretical discussions such as in STS, political ecology, and human geography, practice scholars are beginning to recognise an opportunity to move beyond, or at least contest, the residual human-centrism in theories of practice.

Drawing on the considerable foundations of practice scholarship to date, this collection presents a multidisciplinary exploration of how dynamic non-humans can be (re)conceptualised, represented, and perhaps more fully accounted for in theories of social practice. Through their careful, considered, and at times deliberatively challenging scholarship, contributors draw on a broad range of practice theorists and other intellectual resources, grounding their ideas in detailed empirical work, or connecting their arguments to established practice scholarship. However, there are many more perspectives, approaches, and lines of enquiry to be drawn in exploring the role of dynamic non-humans in theories of social practice, which we hope this collection will inspire.

In writing the chapters for this book, we explicitly asked authors to be provocative and playful with their ideas about non-humans and theories of practice. After all, there is no reason that theory cannot be fun. However, in doing so we were careful not to be dismissive towards the extensive body of work that already exists that can help us make sense of dynamic non-humans. The outcome is reflected here, with chapters that both work within existing theories of practice and draw on other theories and disciplines to make sense of non-human dynamism. The result is, we hope, both entertaining and rigorous, with 'something for everyone' interested in the more-than-human dimensions of social life. Whether your interests concern plants, climate change, robots, or water bottles, we hope this collection inspires further thought-provoking, boundary-testing scholarship on dynamic non-humans and the social practices of which they are a part.

Acknowledgements Writing this introductory chapter, and producing the collection as a whole, has been a joint and equally shared effort between the two editors. To acknowledge this, Cecily Maller is first in the order of editors, while Yolande Strengers is the lead author here. Many thanks to Jenny Atchison, Stanley Blue, and Gordon Walker for gently steering us through tricky territory in writing this chapter. Their comments on previous drafts are greatly appreciated in helping set the scene for this collection.

References

Abram, David. 1997. *The Spell of the Sensuous: Perception and Language in a More-than-human World*. New York: Vintage Books.

Anderson, Ben, and Colin McFarlane. 2011. Assemblage and Geography. *Area* 43 (2): 124–127. https://doi.org/10.1111/j.1475-4762.2011.01004.x.

Barad, Karen. 2003. Posthumanist Performativity: Toward an Understanding of How Matter Comes to Matter. *Signs* 28 (3): 801–831. https://doi.org/10.1086/345321.

Bell, Sarah J., Lesley Instone, and Kathleen J. Mee. In press. Engaged Witnessing: Researching with the More-than-human. *Area*: n/a–n/a. https://doi.org/10.1111/area.12346.

Bennett, Jane. 2010. *Vibrant Matter: A Political Ecology of Things*. Durham, NC: Duke University Press.

Bourdieu, Pierre. 1990. *The Logic of Practice*. Cambridge: Polity Press.

Braun, Bruce, and Sarah Whatmore, eds. 2010a. *Political Matter: Technoscience, Democracy, and Public Life*. Minneapolis, MN: University of Minnesota Press.

———. 2010b. The Stuff of Politics: An Introduction. In *Political Matter: Technoscience, Democracy, and Public Life*, ed. Bruce Braun and Sarah Whatmore, ix–xl. Minneapolis, MN: University of Minnesota Press.

Browne, Alison, Will Medd, and Ben Anderson. 2013. Developing Novel Approaches to Tracking Domestic Water Demand Under Uncertainty—A Reflection on the "Up Scaling" of Social Science Approaches in the United Kingdom. *Water Resources Management* 27 (4): 1013–1035. https://doi.org/10.1007/s11269-012-0117-y.

Castree, Noel. 2003. Environmental Issues: Relational Ontologies and Hybrid Politics. *Progress in Human Geography* 27 (2): 203–211. https://doi.org/10.1191/0309132503ph422pr.

Coole, Diana. 2013. Agentic Capacities and Capacious Historical Materialism: Thinking with New Materialisms in the Political Sciences. *Millennium: Journal of International Studies* 41 (3): 1–19.

Gandy, M. 2003. *Concrete and Clay: Reworking Nature in New York City*. Cambridge, MA: MIT Press.

Giddens, Anthony. 1984. *The Constitution of Society: Outline of the Theory of Structuration*. Cambridge: Polity Press.

Gram-Hanssen, Kirsten. 2009. Standby Consumption in Households Analyzed with a Practice Theory Approach. *Journal of Industrial Ecology* 14 (1): 150–165.

Greenhough, Beth. 2010. Vitalist Geographies: Life and the More-than-Human. In *Taking-Place Non-Representational Theories and Geography*, ed. Ben Anderson and Paul Harrison, 37–54. Farnham: Ashgate Publishing Ltd.

Haraway, Donna J. 1997. *Modest−Witness@Second−Millennium.FemaleMan−Meets−OncoMouse: Feminism and Technoscience*. New York and London: Routledge.

———. 2008. *When Species Meet*. Minneapolis: University of Minnesota Press.

———. 2016. *Staying with the Trouble: Making Kin in the Chthulucene*. Durham, NC: Duke University Press.

Head, Lesley. 2016. *Hope and Grief in the Anthropocene: Re-conceptualising Human-nature Relations*. Oxon, UK: Routledge.

Hitchings, Russell. 2013. Sharing Conventions: Communities of Practice and Thermal Comfort. In *Sustainable Practices: Social Theory and Climate Change*, ed. Elizabeth Shove and Nicola Spurling. Oxon: Routledge.

Houston, Donna, Jean Hillier, Diana MacCallum, Wendy Steele, and Jason Byrne. 2017. Make Kin, Not Cities! Multispecies Entanglements and 'Becoming-World' in Planning Theory. *Planning Theory* 0 (0). https://doi.org/10.1177/1473095216688042.

Ingold, Tim, and Gisli Palsson. 2013. *Biosocial Becomings: Integrating Social and Biological Anthropology*. Cambridge: Cambridge University Press.

Latour, Bruno. 1992. Where are the Missing Masses? The Sociology of a Few Mundane Artifacts. In *Shaping Technology/Building Society: Studies in Sociotechnical Change*, ed. Wiebe E. Bijker and John Law, 225–258. Cambridge, MA: Massachusetts Institute of Technology.

———. 2005. *Reassembling the Social: An Introduction to Actor-network-theory*. New York: Oxford University Press.

Law, John, and John Hassard. 1999. *Actor Network Theory and After*. Wiley.

Lulka, David. 2009. The Residual Humanism of Hybridity: Retaining a Sense of the Earth. *Transactions of the Institute of British Geographers* 34 (3): 378–393. https://doi.org/10.1111/j.1475-5661.2009.00346.x.

Maller, Cecily. 2018. *Healthy Urban Environments: More-than-Human Theories, Routledge Studies in Environment and Health*. Abingdon, UK: Routledge.

Meloni, Maurizio. 2014. How Biology Became Social, and What It Means for Social Theory. *The Sociological Review* 62 (3): 593–614. https://doi.org/10.1111/1467-954X.12151.

Ortner, Sherry B. 1984. Theory in Anthropology Since the Sixties. *Comparative Studies in Society and History* 26 (1): 126–166.

Petersen, Sandra, Susan Houston, Huanying Qin, Corey Tague, and Jill Studley. 2017. The Utilization of Robotic Pets in Dementia Care. *Journal of Alzheimer's Disease* 55: 569–574 http://content.iospress.com/articles/journal-of-alzheimers-disease/jad160703.

Pickering, Andrew. 1995. *The Mangle of Practice: Time, Agency and Science*. Chicago: University of Chicago Press.

———. 2008. Against Human Exceptionalism. Paper presented at a workshop: 'What does it mean to be human?', 25 January. Accessed May 30, 2016. https://ore.exeter.ac.uk/repository/bitstream/handle/10036/18873/XTRwrkshp-250108.pdf?sequence=1&isAllowed=y

Pierce, James, and Eric Paulos. 2010. Materializing Energy. Designing Interactive Systems, Aarhus, Denmark.

Reckwitz, Andreas. 2002a. Toward a Theory of Social Practices: A Development in Culturalist Theorizing. *European Journal of Social Theory* 5 (2): 243–263. https://doi.org/10.1177/13684310222225432.

———. 2002b. The Status of the "Material" in Theories of Culture: From "Social Structure" to "Artefacts". *Journal for the Theory of Social Behaviour* 32 (2): 195–217.

Røpke, Inge. 2009. Theories of Practice—New Inspiration for Ecological Economic Studies on Consumption. *Ecological Economics* 68 (10): 2490–2497.

Rouse, Joseph. 2007. Practice Theory. In *Philosophy of Anthropology and Sociology*, ed. Mark W. Risjord, 639–681. Amsterdam: North-Holland.

Schatzki, Theodore R. 2001. Introduction: Practice Theory. In *The Practice Turn in Contemporary Theory*, ed. Theodore R. Schatzki, Karin Knorr Cetina, and Eike Von Savigny, 1–14. New York: Routledge.

———. 2002. *The Site of the Social: A Philosophical Account of the Constitution of Social Life and Change*. University Park, PA: The Pennsylvania State University Press.

———. 2010. Materiality and Social Life. *Nature and Culture* 5 (2): 123–149. https://doi.org/10.3167/nc.2010.050202.

———. 2016. Practice Theory as Flat Ontology. In *Practice Theory and Research: Exploring the Dynamics of Social Life*, ed. Gert Spaargaren, Don Weenink, and Machiel Lamers, 28–42. Abingdon, Oxon, UK: Routledge.

Shove, Elizabeth. 2003. *Comfort, Cleanliness and Convenience—The Social Organization of Normality*. Edited by Don Slater. New Technologies/New Culture Series. Oxford: Berg.

Shove, Elizabeth, and Mika Pantzar. 2005. Consumers, Producers and Practices: Understanding the Invention and Reinvention of Nordic Walking. *Journal of Consumer Culture* 5 (1): 43–64.

———. 2007. Recruitment and Reproduction: The Careers of and Carriers of Digital Photography and Floorball. *Human Affairs* 17: 154–167.

Shove, Elizabeth, Mika Pantzar, and Matt Watson. 2012. *The Dynamics of Social Practice: Everyday Life and How It Changes*. London: SAGE.

Simonsen, Kirsten. 2012. In Quest of a New Humanism: Embodiment, Experience and Phenomenology as Critical Geography. *Progress in Human Geography* 37 (1): 10–26.

Stengers, Isabelle. 2010. Including Nonhumans in Political Theory: Opening Pandora's Box? In *Political Matter: Technoscience, Democracy, and Public Life*, ed. Bruce Braun and Sarah Whatmore, 3–34. Minneapolis, MN: University of Minnesota Press.

Strengers, Yolande, and Cecily Maller. 2011. Integrating Health, Housing and Energy Policies: The Social Practices of Cooling. *Building Research & Information* 39 (2): 154–168.

Strengers, Yolande, Larissa Nicholls, and Cecily Maller. 2016. Curious Energy Consumers: Humans and Nonhumans in Assemblages of Household Practice. *Journal of Consumer Culture* 16 (3): 761–780. https://doi.org/10.1177/1469540514536194.

Taylor, Brian. 1993. Unconsciousness and Society: The Sociology of Sleep. *International Journal of Politics, Culture, and Society* 6 (3): 463–471.

Walker, Gordon, and Elizabeth Shove. 2007. Ambivalence, Sustainability and the Governance of Socio-Technical Transitions. *Journal of Environmental Policy & Planning* 9 (3): 213–225.

Warde, Alan. 2005. Consumption and Theories of Practice. *Journal of Consumer Culture* 5 (2): 131–153. https://doi.org/10.1177/1469540505053090.

Watson, Matt. 2012. How Theories of Practice Can Inform Transition to a Decarbonised Transport System. *Journal of Transport Geography* 24 (0): 488–496. https://doi.org/10.1016/j.jtrangeo.2012.04.002.

Whatmore, Sarah. 2002. *Hybrid Geographies: Natures Cultures Spaces*. SAGE Publications.

———. 2006. Materialist Returns: Practising Cultural Geography in and for a More-than-human World. *Cultural Geographies* 13 (4): 600–609. https://doi.org/10.1191/1474474006cgj377oa.

Yong, Ed. 2016. *I Contain Multitudes: The Microbes Within Us and a Grander View of Life*. London: The Bodley Head.

Part I

Nature, Materiality and Processes

2

Thriving in the Anthropocene: Understanding Human-Weed Relations and Invasive Plant Management Using Theories of Practice

Jennifer Atchison

Introduction

Weeds are both materially constituted and conceptually defined by their relationships with others, including with people (Harlan and de Wet 1965). Archaeobotany and archaeology tell us of the long history of human-weed relationships, from shifting agriculture at the margins of rubbish dumps, weeds have accompanied humans and benefited from their dispersal across the globe (Mack 2003). This relationship however, cannot be simply understood as linear. As articulated in Vavilov's foundational agronomic work, weeds do not only benefit in the wake of humans. Plants persisting in and around agricultural activities have acted as reservoirs of genetic diversity, hybridising and contributing to the success of many early agricultural crops (Hawkes 2007, 25). In doing so, weeds have affected the spread of crops and people, blurring many of the neatly

J. Atchison (✉)
Australian Centre for Cultural Environmental Research, School of Geography and Sustainable Communities, University of Wollongong, Wollongong, NSW, Australia
e-mail: jennya@uow.edu.au

C. Maller, Y. Strengers (eds.), *Social Practices and Dynamic Non-Humans*,
https://doi.org/10.1007/978-3-319-92189-1_2

cultivated distinctions that might be made around them. To categorise a plant as a weed then is to acknowledge the existence, if not the reciprocity, of the relationships between weedy plants and other living organisms and the extent to which these plants are dynamic and vigorous entities.

Weeds continue to be part of, and present challenges to, the modern agro-industrial complex, but biogeographers and others have noted a significant departure in the recent pattern and scale of invasive plant[1] distributions across the planet (Capinha et al. 2015), where 'gain' from invasive species now exceeds native 'loss' over half of the earth's ice free land (Ellis et al. 2012). In an effort to highlight the significance of this contemporary change and its relationship to other defining features of the Anthropocene, van Kleunen et al. (2015) have labelled this recent departure in the pattern and scale of invasive plant biogeography the 'mass invasion event' (MIE). The problem of invasive species is often considered to be a human one, since their distribution, spread, and the challenges of management also contribute to an understanding of the depth, extent, and limits of human influence (McNeely 2001). But what of the plants themselves? How should we reconcile the discourse of invasive species as indicators or signals of human activity, with an everyday understanding that weeds have impacts and effects? How might we also acknowledge that invasive plants, through their relationships with others, remake the world differently?

In keeping with the aims of this book to provoke discussion and advance conceptualisation of the nonhuman within practice scholarship, in this chapter I consider the capacities of plants and how they might share, express, or confound more human-centred notions of agency. To do this, I use the example of the 'mass invasion event' and human-weed relations in the Anthropocene, including empirical case studies from ethnographic work in Northern Australia, to think through questions of fixity and emergence within practice. Practice scholars from across the broad spectrum of the humanities and social sciences have grappled with

[1] Although the term invasive plant is preferred, in some contexts, including for study participants, the term weed is often used interchangeably. The specific weeds examined in the research study are all recognised by weed scientists to be invasive plants.

questions about the ongoing production of social life and how it is organised. The lens of practice is argued to make possible a view of everyday interactions constituting and consequential in the world (Feldman and Orlikowski 2011, 1241). But, as this volume asks, could this perspective be extended from the human to materials or nonhumans? Do plants, for example, have practices, or do they simply occupy the space of practices which belong to people? How would we recognise the practices of plants, and what difference would this make in practical terms? As radically different living organisms, plants are good at disrupting our human or even our animal-centric view of the worlds we inhabit (Head et al. 2014). For example, as Jenny Rinkinen (2018) (see Chapter 3) illustrates, the biological processes of trees and wood orchestrate and entrain an incumbent set of social practices with particular spatialities and temporalities related to home heating. Our deep and abiding relationship with plants makes them significant in our human lives and worthy of our scholarly attention, but their radical alterity suggests this will not be straightforward (Atchison and Head 2017).

In responding to plants and to the intersections between the social and biological, I draw from Andrew Pickering and Karen Barad and their consideration of the nonhuman in theories of practice, which question the stability of biological organisms. Their emphasis on what *emerges* through practice reorients the frame and consistency of bodies as discrete and as knowable toward what is new and/or different. This reorientation inflects a materialist notion of practice and challenges preconceptions about agency, as well as purist scientific or biosecurity perspectives that invasive species might be known or even managed pre-eminently. In what follows, I briefly outline the recent discourse on the 'mass invasion event' before turning to theories of practice and posthumanist perspectives, which I then illustrate with empirical examples from ethnographic research of invasive plant management.[2]

[2] These examples are drawn from research on the social dimensions of invasive plant management in Northern Australia which included participant observation and semi-structured interviews with government scientists, weed officers, land managers, and Indigenous rangers. This project has been reported elsewhere in Atchison and Head (2013), Atchison (2015), Head et al. (2015), and in Atchison and Head (2017).

The 'Mass Invasion Event'

Although the archaeological and historical evidence is clear that humans have been moving plants around the globe and affecting the earth's vegetation for a long time (Sauer 1947; Harris 1967), sophisticated plant databases now provide a new dimension to this biogeographic story. In late 2015, 38 scientists from across the world published a letter to *Nature*, reporting on the first analysis of the GloNAF (Global Naturalised Alien Flora) database—recording the movement of over 13,000 species across the earth (van Kleunen et al. 2015). The redistribution is complex, and there is intercontinental variation, but the pattern described as MIE upsets the traditional story of 'old world' versus 'new world' biotic exchange, for a more rudimentary north-south 'flip', a pattern of 'donation', predominantly northern to southern hemisphere, with the largest 'donations' from temperate Asia and Europe. This equates to approximately 3.9 per cent of the earth's total vascular flora redistributed (van Kleunen et al. 2015, 100).

The letter to *Nature* builds on a significant body of work which seeks to document and describe changes in the distribution of global plant biodiversity associated with anthropogenic activity, including accelerated rates of species dispersal and transgression of 'natural' biogeographic barriers to dispersal (Capinha et al. 2015). Although there have been biodiversity gains as well as losses (Ellis et al. 2012), and speculation that the processes of human modification may lead to speciation and possibly increasing diversity (Thomas 2013), the overarching pattern is argued to be one of species homogenisation at a global scale attributable to increasing trade and movement of humans and their land uses across the planet (McKinney and Lockwood 1999; Simberloff 2013; Capinha et al. 2015). These changes are correlated with human patterns of activity established since the late Pleistocene (Braje and Erlandson 2013), but are arguably different by function of scale and the diversity of ecosystems involved (Ricciardi 2007; Capinha et al. 2015) and where related factors including climate change have facilitated establishment (Chown et al. 2015). Together with this body of work, van Kleunen et al.'s article is significant in drawing attention to something empirically new regarding important biodiversity variables and implications for ecological systems. As the authors point out, this information should prompt integrated efforts to understand and respond to invasive plants (van Kleunen et al. 2015). Here though, I want to draw attention to the attribution of agency within the scientific discourse of the MIE.

The MIE illustrates a persistent tension in the attribution of agency within the hard or purist ecological sciences with regard to the nonhuman. This is a related but differently articulated binary to the native/alien, nature/culture polarity that social scientists have critiqued extensively (Warren 2007). Within the MIE discourse, there is a consistent view that the factors that contribute to the spread and success of invasive plants, such as the erosion of biogeographic barriers through global trade, are consistent with and defining features of the Anthropocene (van Kleunen et al. 2015), that is, the MIE is also described as human mediated and a signal of human modification. Humans are strong 'shapers' in this global redistribution, where the arteries of exchange are human global trade and migration routes (van Kleunen et al. 2015). Plant distribution is viewed as 'a result of human activity', with the implicit understanding that this is problematic. Plants are described as passive, they merely 'accumulate' or are accumulated (by continents) (van Kleunen et al. 2015).

On the other hand, invasive plants are also clearly understood by invasive species biologists and cognate scientists to have capacities to act in the world, including apart from humans. Indeed, their categorisation within these disciplines as 'invasive' signals this capacity through both mobile and displacing activities. For example, in their 2010 review, Pyšek and Richardson (2010, 27) ask: 'Which species invade?', 'Which habitats are invaded?', and 'How can we manage invasions?'. These questions draw attention to the seemingly unquestionable and active quality of invasive species, as well as to the human capacity to act back at them through management action, richly illustrating the depth ascribed to the power of invasive species. These capacities also make invasive plants problematic precisely because they are at arm's length from the human; they 'escape' and act in the world outside, beyond or indifferent to human influence, as well as with, or in association with, humans.

Notwithstanding the obvious and problematic anthropomorphism associated with describing the agency of the nonhuman world, there are contradictions and complexities apparent in articulating amassed and changing nonhuman life as part of increasing anthropogenic influence across the earth. This is the provocation of attending to relationships with humans, but recognising the irreducibility of the nonhuman component of the 'disaster' (Clark et al. 2013). Drawing on both Clark and Yusoff, who call for 'an expanded sense of the nonhuman in scope and scale', Castree (2014, 456) has argued that the Anthropocene demands more sophisticated ontological perspectives—new intellectual tools to deal with the imbricated nature of human-nature relationships and the complex problems emerging. This might include new, and perhaps less muscular, sensibilities, for example, which better attune us to both the more-than-human and to the 'inhuman world' indifferent or even hostile to our existence (Castree 2014, 456). Further, the limitations of approaches based on purist science that continue to delimit 'natural' nature from anthropogenic culture (Castree 2014) are not merely semantic. As the examples I illustrate below demonstrate, the way we attend to these relationships has practical implications.

Questioning the Stability of Bio-organisms and Material Agency

Given the body of work well underway within human geography and elsewhere to recognise and understand the agency of the more-than-human world (Whatmore 2006; Panelli 2010), it is worth reiterating that theories or accounts of nonhuman agency have not always been acknowledged or accepted and, in some circles, remain contentious or, at best, peripheral (Castree et al. 2004). The substantial body of post-humanist work describing hybridity, networks, and assemblages has provided materially rich, detailed, and challenging accounts of life in the world. It is challenged, however, by the terms and frames of reference rooted in what it means to be human (Lulka 2009). Whether existing frames can be applied or new ones need to be conjured is an ongoing question. See, for example, Kohn's (2013) provoking 'anthropology beyond the human'. In this context, it is also useful to remember that plants in particular have mostly been backgrounded and are usually conceptualised in relation to animals (Head et al. 2015), for example, where they are described as passive, which runs counter to activity—one of the core constructs of social thought (Schatzki 2007). Likewise, human-plant practices are often taken for granted in that they may involve everyday or seemingly mundane activities (Phillips 2016). To follow seriously then, the assertion that plants take part actively in the world, is never straightforward and certainly not central to the contemporary concerns of invasive species science.

Here, I look toward the new materialisms and focus on nonhuman agency in theories of practice. In this case, I want to highlight other constructive ways of thinking about invasive plants beyond the dominant traditions of invasive species biology and purist ecology. For the most part, practice scholars such as Bourdieu and Giddens, amongst others, have focused on and indeed confined their studies of practice to the human, elaborating the ways in which human practice and agency is generative of structure, norms, and order (Feldman and Orlikowski 2011). There are other practice scholars though, from where this volume departs, who argue that materials and dynamic nonhumans are not just the stuff that social human beings organise but entities who are also consequential and significant in the formation of social life. Whether or not nonhumans perform practices in their own right, however, are ongoing concerns.

As a sociologist and scholar of science and technology studies, Andrew Pickering is most well-known for his accounts of the practice of science beyond a Descartian representation of objective reality and in encounters between humans and materials (Pickering 1995). Pickering's previous work as a physicist provided richly detailed examples to draw upon, such as the development of the bubble or experimental particle chamber (Pickering 1995), in developing his theories of material agency in the practice of science. He also made the case and indeed took his ideas into the nonhuman world, articulating examples of more lively or living organisms, the environment, and its relationship to people through practice (Pickering 2005). Although Pickering also developed theories of human agency, it is his discussion of material or nonhuman agency, in relation to human agency, and as different from it, that is of most relevance here. In *The Mangle of Practice* (1993, 562), Pickering specifically locates agency amongst the human and the material or nonhuman, that is, 'agency is an entanglement of constitutive human and nonhuman elements that make up the mangle'. The mangle is a dance of agency or a 'dialectic of resistance and accommodation' between different agencies, reciprocal and intertwined. Taking inspiration from foundational work by Callon, Latour, and Law in relation to actor-network theory (ANT) and the assertion that agency can also reside with the material or nonhuman (Schatzki 2001), there are significant parallels in Pickering's approach and ANT. Although he diverges in insisting that human and material or nonhuman agency are qualitatively different and not interchangeable, it is his articulation of *temporal emergence* and *constitutive intertwining* that stand out (Pickering 1995, 15).

In contrast to social constructivist arguments, where the social is considered to be an independent variable, Pickering argues what is also 'at stake' in the mangle of practice are the new social becomings or 'coupled becomings of the human and nonhuman' (Pickering 2005, 35). In other words, the 'social' is not an independent variable, but comes into being as an assemblage emerges. Alecia Jackson (2013, 743) notes this as a mutual production, or mutual constitution of the human and nonhuman. Pickering's insistence that agency is temporally emergent was further developed in his 2005 essay on Asian eels and global warming (Pickering 2005). Although the first generation of eels may have been *temporally prior*, this new assemblage or object that developed over time, Pickering argues, was a new 'heterogeneous assemblage' that could not have been predicted or known in advance before the eels arrived (Pickering 2005, 34). In other words, the establishment of Asian eels in the USA as an assemblage takes on its own form and character and social life due to unique place and time contingencies.

Pickering has been critiqued from various angles, but the most notable is from Barad (2003, 807) who credits his account of nonhuman agency, but argues his invocation 'takes for granted the humanist notion of agency as a property of the *individual*'. Also trained as a physicist, Barad's intellectual project attempts to wrestle with the confounding and transformational ideas presented by quantum physics, and her insistence that any account of matter must deal with all matter (human and nonhuman), describing the relationships between discursive practice and material phenomena. Her challenging theory of intra-action makes the ontological proposition that matter *also* emerges, including discursively. This irreducibility or inseparability of matter and concern, she argues, also obligates a redistribution of responsibility (Barad 2007).

For me, it is not insignificant that one of Pickering's original posthuman objects is an invasive species assemblage; there is purchase in his notion that the translocation and haphazard establishment of Asian eels in the USA enables the spatial and temporal contingencies involved to be identified. However, I think we can also make more of this example, in relation to constitutive intertwining and the possibility that organisms themselves, or at least the way we 'cut', define, or measure them, are constituted through practice (Barad 2003). Invasive plant management practices are traditionally conceptualised as boundary-making practices, performed by people attempting to articulate and stabilise the demarcation or separation between particular types of plants, people, and places. In this case, while following Barad's challenging propositions through is not straightforward, practice might become a potentially productive space in which the boundaries and new becomings of the human and the nonhuman might also usefully be examined.

To illustrate how attention to practice might enable such a shift in thinking about materiality and agency, I outline three empirical examples drawn from ethnographic work across Northern Australia of invasive plants in natural resource management practices. In significant ways, our view of plants is always partial (Atchison and Head 2017) however, invasive plants do facilitate a more tangible and active sense of the contours of plant agency. The practices described below are sometimes formalised or prescribed, and sometimes not. Although the MIE is mostly characterised in the literature as hegemonic and global, the ethnographic examples illustrated here emphasise the micro scale and multiple constitutions of human and nonhuman, fleshing out the various humans and plants as they perform practices together.

The Intimate and Embodied Practices of Invasive Plant Management

Getting a Feeling for *Mimosa*

Species from the genus *Mimosa* are widely known as 'touch sensitive' plants because they respond to stimuli (Braam 2005), folding their leaves when touched. These responses are of great interest to scientists because they highlight plant capacities previously undescribed in ways that are easily observable by humans. They also serve to highlight the limitations of any theoretical attempts to separate the social from the biological, or the functions of organisms in relation to other beings or to the nonliving world. As biological scientist Braam (2005, 385–386) contends, 'understanding the mechanisms of perception and response are likely fundamental to understanding plant biology'.

To describe *Mimosa pigra* from ecological science, it is a transformer species (Grice 2006) and a 'Weed of National Significance' in the Australian national biodiversity governance framework. It invades undisturbed wetland environments and spreads by water and the movement of wetland animals and birds dispersing seeds. Mimosa rapidly outcompetes other vegetation, forming monocultural stands that quickly change the hydrological and ecological function of wetlands. Understanding *Mimosa pigra* responses is a key part of the work of biocontrol scientists Ben, Stephanie, and Nadia, who grow it for the national mimosa biocontrol programme at a Northern Territory government facility near Darwin, Australia. Biocontrol is a set of experimental practices between people, plants, and insects whereby control of an invasive plant might emerge, but cannot be imposed (Atchison 2015). The biocontrol team in Darwin is attempting to understand how mimosa responds and survives under controlled insect herbivory stress. Stephanie explained:

> So a lot of what Ben has done is just getting that feel for how a weed grows and if you're trying to work an insect on it, what makes it work the best.
>
> Interviewer
>
> And how do you do that? How do you get a feel for a weed?

> Well, getting a feel is probably strange because every season is different so looking at a weed and how it appears this season might be different to last season and in different locations it's also going to be different. So you're collecting data over, as long a period as possible so that you can look for trends and shapes, patterns in the data and see what's sort of typical and what the boundaries are, you know, what the extremes are and what you can expect in terms of, how quickly it grows, how much seed it churns out and how it spreads, how the insects grow and how they spread and what things have knocked them or set them back and what seems to be conducive to their survival.
>
> Stephanie (Technical officer, NT)

For the past ten years, the team has been attuning themselves to mimosa—to its extremes, its patterns, its inconsistencies, its variations and associations—by growing it. This is an embodied set of practices (Doody et al. 2014)—Ben selects each seed with a pair of fine tweezers, inserts it into a vermiculite growing medium, and then turns them out into neat plastic boxes which are watered and placed under grow lights in the temperature- and light-controlled cupboards (Fig. 2.1). The seedlings

Fig. 2.1 Inspecting the growing *Mimosa pigra* seedlings in the biocontrol laboratory. Photo © J. Atchison

are attended to on a daily roster, their growth features observed and noted. In learning how to kill mimosa, Ben must first tend and care for it.

> Even though our goal is to control the weeds because we're feeding our insects… you're also trying to look after your plants to make them the best possible quality to feed your insects, and so you do have to still care for your captive weeds. Which means that you need to understand the weed.
> Ben (Technical officer, NT)

> Because we understand it a lot more and we get to go on a flood plain and look at this weed and you see this plant growing in different environments and you think, wow this is beautiful mimosa, this is awesome. So we do end up viewing some of them from a different perspective.
> Nadia (Technical officer, NT)

In contrast to chemical control or management, the scientific practices of biocontrol described above include close observation and analysis as well as growing and caring for mimosa. Biocontrol practices then can be thought of as both social and biological processes of attuning people and insects to mimosa cycles, time frames, and contingencies. And mimosa is not blank or passive matter here—it emerges as spatially and temporally variable, sick, healthy, or resilient within these practices. Although not complacent about the impacts, these scientists have developed respect for an invasive plant and, at a personal level at least, are able to discuss the ways it affects, excites, or compels them. Although in many ways human and partial, their perspective of mimosa and its agency is informed and shaped by biocontrol practices.

Mowing the (Gamba) Grass

Elsewhere in Darwin at a small recreation reserve, Brad manages a community Landcare programme. Once used as a dumping site, the reserve and the freshwater wetland lagoon it encompasses is now surrounded in patch-work fashion by small peri-urban landholdings with newly planted gardens amongst the savanna eucalypt forest. Gamba grass (*Andropogon gayanus*), another 'Weed of National Significance', was introduced to Northern Australia in the 1930s and is part of a longer history of colonising the country through pasture production and the naturalisation of exotic species (Cook and Dias 2006). Gamba is also an ecosystem transformer. In recent decades, gamba has proliferated and spread across Darwin and surrounding areas where it fuels intensely hot fires that reach up into the tree canopy and result in tree death (Rossiter et al. 2003).

The gamba that is present might be controlled through chemical application and spraying practices, but Brad cannot spray because herbicides such as glyphosate are prohibited from use around sensitive aquatic environments—a regulatory measure used to protect birds, frogs, fish, and reptiles. Brad previously mowed the gamba on the reserve with a push mower by hand, but with the money from a small community grant, he has now purchased a small tractor and slasher, a ride-on mower, and a set of 'whipper snippers' (Fig. 2.2). Each year and sometimes more than once a year, Brad mows a wide strip around the lagoon to create a fire break.

Fig. 2.2 One of the community-owned mowers used to manage *Andropogon gayanus* and other grasses. Photo © J. Atchison

He also mows along the walkways and picnic area of the reserve. This doesn't kill the gamba, but keeps it, as he described it, 'tidy and clean'.

Elsewhere around Darwin, other people are also mowing gamba grass. On a nearby mango farm, an orchardist mows the gamba amongst the fruit tree crop. Again, the chemical spray used elsewhere is not able to be used around food for human consumption. The gamba doesn't die but is maintained, in a short clipped, if somewhat clumpy, dry 'lawn'. Around the city airport also, gamba is mown rather than sprayed in order to reduce fuel loads around the aviation infrastructure and fuel stores. Gamba grass responds to mowing depending on its growth phase. Mown early in the year, its growth slows. Mown later in the year after seeding, mowing can have the effect of spreading the seed.

Managing a grassy weed such as gamba grass, whose tiny seeds can be blown in the wind, is fraught with logistical and practical complexities, and management practices do not always implicate clear spatial demarcations. Gamba prevaricates between the vagaries of seasonality and human management ideals. Even where land managers are convinced of the need to manage gamba, this weed illuminates the gaps in rigid regimes where chemical control is not possible, or where mowing machines have been unable to reach along fence lines or in ditches. People are responding to gamba but not always in ways conditioned by clear management terms or with outcomes that are predictable. Conditioned by context, as well as the seasonal growth of the grass itself, people tend to gamba in ways that both limit and contribute to its proliferation.

Washing the Rubber Vine

The final example comes from Northern Queensland and the town of Mount Surprise, where bright orange road signs command the attention of visitors. The first instructs travellers to 'Prevent the spread of weed seed', before they slow to the designated 60 kilometre speed limitation zone. Through the town, immanent arrival at the weed washdown facility is signalled at 100 metre intervals. My colleague and I patiently wait in the queue of vehicles, paying careful attention to close windows and fold rear-vision mirrors, as instructed by the automated sound system. We drive across the metal grate and park, while high-pressure water hoses, attached to a timer, make themselves busy on the mud and hidden seeds attached to our vehicle. The hoses jettison a regulation amount of water which collects in a holding tank, before it spills along a pipe and flows out into an adjacent paddock. Two further smaller signs at the exit thank us as we leave; unexpectedly, we have become weed hygiene practitioners.

Mount Surprise is currently within the boundaries of the national control zone for rubber vine (*Cryptostegia grandiflora*), an area demarcating abundant infestations to the west, from more localised or occasional distributions to the east. A 'Weed of National Significance', rubber vine is described as a vigorous climbing plant which threatens both biodiversity and grazing enterprises (DAFF 2012) and is the subject of strict state and national biosecurity governance. Within the control zone, the management focus is on asset protection, rather than eradication, due to the scale and density of the incursion.

The humble washdown facility now in place provides insights into the constellation of management issues and practices coalescing around any single weed management initiative; many observers agree that hygiene of machinery and vehicles is key, but there are debates how best to do this and who should pay. Positioning washdown facilities on one side of the road means that alternative directional movements are missed or traffic flows may be disrupted. The establishment cost of a washdown facility can easily run to over 100,000 dollars (AUD)—well beyond the means of most local government agencies. While there is a case that such facilities provide public benefit, the network of roads in Northern Australia is extensive and in places disproportionately used by private industry. And, where facilities do exist, it is not always clear who should be required to use them or how this would be enforced. Weed washdown requires a water supply, power, and people who can periodically inspect for seeds sprouting in the flood-out area, as well as carry out maintenance. As one weed officer explained, he'd had 'enormous trouble' getting council bureaucrats to contribute non-financial support for the project. After much negotiation they had settled on a location, but he conceded that the facility was now probably quite ineffective because the land was not able to be secured at the ideal location further east.

Although rubber vine management plans and treatment manuals depict clearly sectioned geographic boundaries, delineating and limiting rubber vine also involves a myriad of other activities—writing grants, getting permissions, and paperwork. This other kind of 'weed work' is vital to the success of much on-ground action, but it requires extensive (human) networks and government contacts as well as considerable travel. Like other routine and domestic duties, these other practices of weed work are not readily accounted for and it is difficult to make them visible, but they underpin much environmental and biosecurity management across the country. Managing weeds has become imbricated, even procedural or routine, in the daily lives of people in subtle but no less significant ways. Rubber vine in its material form is seemingly absent, as if it might actually be 'washed away', and yet the ever-present threat of small seeds and their anticipated mobility and fecundity brings into being and conditions the specific ways in which people respond.

Conclusion

What is done in the 'mangle of practice' or in the 'dialectic of resistance and accommodation'? How do we conceive of and adequately describe or account for nonhuman or 'nonstandard' agency (Pickering 1995) in theories of practice? In this chapter I have sought to illustrate how practices and the particular perspectives of Pickering and Barad might be used to open up questions of agency beyond the human and also inform environmental management in the context of invasive plant management and the mass invasion event. Within biosecurity, invasive plant management practices are traditionally depicted as management actions performed by people. Instead, as I have shown, invasive plants not only respond, and influence what people do, they do so in ways that are not always easily predicted or imagined by the humans involved, with the effect of reshaping the practices of invasive plant management. In some cases, the plants defy the boundaries that are drawn around them, or at least it is possible to see that the demarcations human envisages cannot be so firmly fixed. As ecosystem transformers, it is clear also that these particular plants form relations, beyond those with people, and are consequential in the world. These brief examples quickly confound and disrupt any notion that what happens in the mangle of practice is only the result or effect of something solely reducible to human action.

What difference might it make to consider the specific agencies and dynamism of plants, including in practical terms? There are at least two reasons why attending to practice, theorised from both Pickering's and Barad's perspectives, might be useful for thinking about the MIE and invasive plant management more broadly. First, Pickering (2005) cautions that risk is a symptom of scientifically mediated posthumanism; the assemblage of risk is both produced and mediated by science. There is an inevitable and inexorable quality to the biological and ecological literature on invasive plants and the MIE, especially at a global scale, often implicating rigid, universalised, and non-flexible approaches to thinking about and managing change. The idea of temporal emergence, however, enables assumptions about the *a priori* stability, or predictability, of organisms and their agency to be questioned. Invasive species become risky or problematic in and through their intra-actions with others—these intra-actions and outcomes cannot always be known or predicted. The examples I have illustrated suggest that accounts of invasive species are not only about homogeneity and loss of diversity, significant though these are. The distinction is subtle, but might practically reorient thinking about how people plan to live in and amongst invasive species and manage their incumbent risks. Barker's (2008) analysis of the active role of gorse (*Ulex* spp.) in emergent biosecurity responses is also exemplary in this regard.

Second, this approach draws attention to what is both materially and discursively new. Scientists, at least, are alert to the combinations of what is different in a biogeographic sense, but the implication of a focus on practice is also that there are new socio-material worlds being produced. To draw from both Pickering and Barad, much less attention has been paid to what is also concurrently *at stake* for us, and what or how humans are becoming in the MIE, with a sense that humans and plants are also co-constituted in new ways through being entangled in specific management practices. The history of plant domestication and agricultural production documents how people, plants, and others condition and are conditioned by the practices of living and dying together.

In this sense, contemporary human relationships with weeds continue a long history of coproduction and a continuation of global-scale change. However, the challenges presented by the current pattern and scale of plant 'invasion' suggest there are new ways of being in the world, and new practices emerging between people and plants that I suggest provide compelling reasons to attend to and situate plants in the Anthropocene. Certainly, invasive species management is pervasive and has become an overarching theme of contemporary natural resource agendas. To what extent weed management framed in particular ways, such as those emphasising killing and eradication is normal or becomes normalised, is an important question for future relationships with nature.

Scientific accounts of invasive plants are significant in alerting us to the biogeographic changes underway across the earth's surface. The rhetoric of control and/or winnable 'wars' against invasive species, however, does nothing to address the logistical realities of having to live with invasive life on practical terms. Similarly, the policy rhetoric of invasive species management can sometimes have the effect of disconnecting the work and labour of invasive plant management from the embodied and mediated interactions it demands where, for example, failures to control invasive plants become the failures of people to understand or apply what is understood to be 'best practice'. By shifting the temporality and scale of inquiry, attention to social practices challenges purist ecological and biosecurity perspectives that invasive plants might be known in advance of their arrival and reduced simplistically, to threats for straightforward eradication. In doing so, it provides the scope to move beyond an imagined world without weeds and attend more carefully to the world of invasive life that is emerging.

Acknowledgements The research reported here was funded by an Australian Research Council Laureate Fellowship to Professor Lesley Head (FL0992397). I thank Lesley and Dr Catherine Phillips for always productive conversations about plants, Stephanie Toole for assistance in the field, and colleagues at AUSCCER for engaging discussions on theories of practice.

References

Atchison, Jennifer. 2015. Experiments in Co-existence: The Science and Practices of Biocontrol in Invasive Species Management. *Environment and Planning A* 47 (8): 1697–1712.

Atchison, Jennifer, and Lesley Head. 2013. Eradicating Bodies in Invasive Plant Management. *Environment and Planning D: Society and Space* 31 (6): 951–968.

———. 2017. Rethinking Ethnobotany? A Methodological Reflection on Human-Plant Research. In *Participatory Research in More-than-Human Worlds*, ed. Michelle Bastian, Owain Jones, Niamh Moore, and Emma Roe, 178–191. London: Routledge.

Barad, Karen. 2003. Posthumanist Performativity: Toward an Understanding of How Matter Comes to Matter. *Signs: Journal of Women in Culture and Society* 28 (3): 801–831.

———. 2007. *Meeting the Universe Halfway: Quantum Physics and the Entanglement of Matter and Meaning*. Durham, NC: Duke University Press.

Barker, Kezia. 2008. Flexible Boundaries in Biosecurity: Accommodating Gorse in Aotearoa New Zealand. *Environment and Planning A* 40 (7): 1598–1614.

Braam, Janet. 2005. In Touch: Plant Responses to Mechanical Stimuli. *New Phytologist* 165 (2): 373–389.

Braje, Todd J., and Jon M. Erlandson. 2013. Human Acceleration of Animal and Plant Extinctions: A Late Pleistocene, Holocene, and Anthropocene Continuum. *Anthropocene* 4: 14–23.

Capinha, César, Franz Essl, Hanno Seebens, Dietmar Moser, and Henrique Miguel Pereira. 2015. The Dispersal of Alien Species Redefines Biogeography in the Anthropocene. *Science* 348 (6240): 1248–1251.

Castree, Noel. 2014. Geography and the Anthropocene II: Current Contributions. *Geography Compass* 8 (7): 450–463.

Castree, Noel, Catherine Nash, Neil Badmington, Bruce Braun, Jonathon Murdoch, and Sarah Whatmore. 2004. Mapping Posthumanism: An Exchange. *Environment and Planning A* 36 (8): 1341–1363.

Chown, Steven L., Kathryn A. Hodgins, Philippa C. Griffin, John G. Oakeshott, Margaret Byrne, and Ary A. Hoffmann. 2015. Biological Invasions, Climate Change and Genomics. *Evolutionary Applications* 8 (1): 23–46.

Clark, Nigel, Vasudha Chhotray, and Roger Few. 2013. Global Justice and Disasters. *The Geographical Journal* 179 (2): 105–113.

Cook, Garry D., and Lesley Dias. 2006. It Was No Accident: Deliberate Plant Introductions by Australian Government Agencies During the 20th Century. *Australian Journal of Botany* 54 (7): 601–625.

Department of Agriculture and Fisheries Queensland (DAFF). 2012. Rubber Vine *Cryptostegia grandiflora* and *Cryptostegia madagascariensis*. Fact Sheet. https://www.daf.qld.gov.au/__data/assets/pdf_file/0020/52544/IPA-Rubber-Vine-PP11.pdf.

Doody, Brendan J., Harvey C. Perkins, Jon J. Sullivan, Colin D. Meurk, and Glenn H. Stewart. 2014. Performing Weeds: Gardening, Plant Agencies and Urban Plant Conservation. *Geoforum* 56: 124–136.

Ellis, Erle C., Erica C. Antill, and Holga Kreft. 2012. All is Not Loss: Plant Biodiversity in the Anthropocene. *PLoS One* 7 (1): e30535.

Feldman, Martha S., and Wanda J. Orlikowski. 2011. Theorizing Practice and Practicing Theory. *Organization Science* 22 (5): 1240–1253.

Grice, Anthony C. 2006. The Impacts of Invasive Plant Species on the Biodiversity of Australian Rangelands. *The Rangeland Journal* 28 (1): 27–35.

Harlan, Jack R., and Jan M.J. de Wet. 1965. Some Thoughts About Weeds. *Economic Botany* 19 (1): 16–24.

Harris, David R. 1967. New Light on Plant Domestication and the Origins of Agriculture: A Review. *Geographical Review* 57 (1): 90–107.

Hawkes, Jack G. 2007. The Ecological Background of Plant Domestication. In *The Domestication and Exploitation of Plants and Animals*, ed. Peter J. Ucko and Geoffrey W. Dimbleby, 17–29. New Brunswick, NJ: Transaction Publishers.

Head, Lesley, Jennifer Atchison, and Catherine Phillips. 2015. The Distinctive Capacities of Plants: Re-thinking Difference Via Invasive Species. *Transactions of the Institute of British Geographers* 40 (3): 399–413.

Head, Lesley, Jennifer Atchison, Catherine Phillips, and Kathleen Buckingham. 2014. Vegetal Politics: Belonging, Practices and Places. *Social and Cultural Geography* 15 (8): 861–870.

Jackson, Alecia Y. 2013. Posthumanist Data Analysis of Mangling Practices. *International Journal of Qualitative Studies in Education* 26 (6): 741–748.

Kohn, Eduardo. 2013. *How Forests Think: Toward an Anthropology Beyond the Human*. Berkley, CA: University of California Press.

Lulka, David. 2009. The Residual Humanism of Hybridity: Retaining a Sense of the Earth. *Transactions of the Institute of British Geographers* 34 (3): 378–393.

Mack, Richard N. 2003. Global Plant Dispersal, Naturalization, and Invasion: Pathways, Modes, and Circumstances. In *Invasive Species: Vectors and Management Strategies*, ed. Gregory Ruiz and James T. Carlton, 3–30. Washington, DC: Island Press.

McKinney, Michael L., and Julie L. Lockwood. 1999. Biotic Homogenization: A Few Winners Replacing Many Losers in the Next Mass Extinction. *Trends in Ecology & Evolution* 14 (11): 450–453.
McNeely, Jeffrey A. 2001. *The Great Reshuffling: Human Dimensions of Invasive Alien Species*. Cambridge, UK: IUCN.
Panelli, Ruth. 2010. More-than-human Social Geographies: Posthuman and Other Possibilities. *Progress in Human Geography* 34 (1): 79–87.
Phillips, Catherine. 2016. *Saving More Than Seeds: Practices and Politics of Seed Saving*. London: Routledge.
Pickering, Andrew. 1993. The Mangle of Practice: Agency and Emergence in the Sociology of Science. *American Journal of Sociology* 99: 559–589.
———. 1995. *The Mangle of Practice. Time, Agency and Science*. London: University of Chicago Press.
———. 2005. Asian Eels and Global Warming: A Posthumanist Perspective on Society and the Environment. *Ethics & The Environment* 10 (2): 29–43.
Pyšek, Petr, and David M. Richardson. 2010. Invasive Species, Environmental Change and Management, and Health. *Annual Review of Environment and Resources* 35: 25–55.
Ricciardi, Anthony. 2007. Are Modern Biological Invasions an Unprecedented form of Global Change? *Conservation Biology* 21 (2): 329–336.
Rinkinen, Jenny. 2018. Seeing Wood for the Trees: Placing Biological Processes Within Practices of Heating and Harvesting. In *Social Practices and Dynamic Non-humans: Nature, Materials and Technologies*, ed. Cecily Maller and Yolande Strengers, 47–62. Cham, Switzerland: Palgrave Macmillan.
Rossiter, Natalie A., Samantha A. Setterfield, Michael M. Douglas, and Lindsay B. Hutley. 2003. Testing the Grass-Fire Cycle: Alien Grass Invasion in the Tropical Savannas of Northern Australia. *Diversity and Distributions* 9 (3): 169–176.
Sauer, Carl O. 1947. Early Relations of Man to Plants. *Geographical Review* 37 (1): 1–25.
Schatzki, Theodore. 2001. Introduction: Practice Theory. In *The Practice Turn in Contemporary Theory*, ed. Theodore Schatzki, Karin K. Cetina, and Eike V. Savigny, 10–23. London: Routledge.
———. 2007. Introduction. *Human Affairs* 17: 97–100.
Simberloff, Daniel. 2013. Commentary: Biological Invasions: Prospects for Slowing a Major Global Change. *Elementa Science of the Anthropocene* 1: 8.
Thomas, Chris D. 2013. The Anthropocene Could Raise Biological Diversity. *Nature* 502 (7469): 7.

van Kleunen, Mark, Wayne Dawson, Franz Essl, Jan Pergl, Marten Winter, Ewald Weber, Holger Kreft, et al. 2015. Global Exchange and Accumulation of Non-native Plants. *Nature* 525: 100–103.

Warren, Charles R. 2007. Perspectives on the Alien 'Versus Native' Species Debate: A Critique of Concepts, Language and Practice. *Progress in Human Geography* 31 (4): 427–446.

Whatmore, Sarah. 2006. Materialist Returns: Practising Cultural Geography in and for a More-than-Human World. *Cultural Geographies* 13 (4): 600–609.

3

Seeing Wood for the Trees: Placing Biological Processes Within Practices of Heating and Harvesting

Jenny Rinkinen

Introduction

It is now widely acknowledged that materiality is crucial for all social practices. By definition, a social practice exists as an entity which draws upon a set of material resources when it is carried out—such as food and energy for cooking and eating or water for showering (Shove et al. 2012). However, within material culture studies and the sociology of consumption, what is understood by 'materiality' is often limited: materiality is seen as static and discrete 'objects', as formable things that are used in the process of consumption (Miller 1998; Warde 2005). As has been argued, this emphasis on materiality that prioritises finished artefacts over the

---

This chapter develops some of the ideas presented in my PhD Thesis, entitled 'Demanding energy in everyday life: Insights from wood heating into theories of social practice' (Rinkinen 2015).

J. Rinkinen (✉)
Department of Sociology, DEMAND Centre (Dynamics of Energy, Mobility and Demand), Lancaster University, Lancaster, UK
e-mail: jenny.rinkinen@lancaster.ac.uk

C. Maller, Y. Strengers (eds.), *Social Practices and Dynamic Non-Humans*,
https://doi.org/10.1007/978-3-319-92189-1_3

properties of materials overlooks their enrolment in dynamic form-making processes (Ingold 2013).

Recent advances in theories of social practice have pushed for a more extended view of materiality that covers resources and complex arrangements such as systems of provision and also recognises how they stretch within and between social practices (e.g., Shove et al. 2015; Shove 2017; Morley 2016). In Shove's (2017) account, 'resources', including forms of energy, are materials that are used up in the doing of practice. They differ from the more static 'infrastructural arrangements' and systems of provision and the more actively used and moulded 'things' of practice. Infrastructural, device-oriented, and resource-based relations are thoroughly inseparable, as they are welded together in various combinations across different practices (Shove 2017). Schatzki also acknowledges the wide range of material arrangements, which for him include 'humans, artefacts, organisms, and things of nature' (Schatzki 2010, 129). 'Natural' materiality is something Schatzki distinguishes as happening or changing on its own, often subject to principles or laws *not* of human making (Schatzki 2010).

Even though these theoretical accounts recognise the complex, fluid, and dynamic nature of materiality, it remains the case that in analyses of practice, great emphasis is still placed on the artefactual, human-made domain of materiality, ignoring and overlooking the achievements and impacts of non-humans and 'living organisms'. The conceptualisation of non-human entities such as wood, oil, or even food as static resources or things in use has led to a poor understanding of materials that give rise to effects in the world and are fundamentally constitutive of things and practices. For example, the lives of biological non-humans affect the trajectories of practices by developing capabilities as part of ecosystems, running out, or 'entering' and 'exiting' practices and thus leaving their traces. In overlooking the biological dimensions of social practices, there is the risk that potential processes of change are overlooked. This is worrying if we want to understand social practices as part of and as embedded within the complex processes of nature.

Drawing inspiration from ecological anthropologists, this chapter aims to move towards 'biological' understandings of materiality within theories of practice. It does so by raising questions about the challenges of harnessing biological ideas for thinking about practices: why should it be done and how? There has been a recent interest in how the human body is inserted into practices (e.g., Wallenborn and Wilhite 2014), but my interest in the biological is limited to processes, such as the growth and decay, of living non-human organisms.

The urge to widen the understanding of materiality stems from the notion that non-human matter is living, organic, and dynamic—and thoroughly and crucially intertwined with social practices. In some respects, all matter is already natural as it is difficult to draw a line between what is natural and what is not, with many materials classified as hybrids composed of organisms and machines (Haraway 2013). Yet, theoretically, it is not clear how the biological processes embedded in all inert materials are part of social practices. How do biological processes and social processes intertwine, co-evolve, and change practices?

Following this idea that both natural and social processes are part of an intertwined process of becoming, I take up a suggestion that stems from ecological anthropology to turn from the 'objectness' of things to the material flows and formative processes wherein they come into being (Ingold 2013). In doing so I take the understanding of materiality in theories of practice further, not only considering living organisms as 'resources', which are the tangible, formable matter used in the reproduction of practices (Shove 2017), but evoking more sensitivity to the more intangible biological and natural processes of resources. As Atchison (2018) also argues, non-humans such as plants are active participants within a variety of sets of practices. I suggest that materiality, conceptualised as a constant process of becoming, mediates standardisation and localisation, as well as the transformation and rooting of practices.

In reflecting on these ideas, I turn to a case of small-scale wood-based space heating and harvesting practices in Finland and depict a common resource—wood—as a living organism and as a distinctive material element of practice. Small-scale wood heating has long roots in different countries, but it is particularly popular in the Nordic countries and in Finland. In past research with colleagues, I have explored questions of time-use and temporal patterning of consumption and technological solutions as they relate to wood-heating social practices (Rinkinen 2015; Jalas and Rinkinen 2016), and distinguished between the various ways in which people engage with material things in keeping warm in winter (Rinkinen et al. 2015). However, questions remain about the 'livingness' of wood, which is commonly downplayed or unacknowledged when its role is reduced to a 'material' resource in accounts of practice: the notion that wood once was a *tree*, a living entity, is forgotten. In this sense, talking about 'wood' as a resource conceals its lively features. In this chapter, I use the case of wood harvesting and heating to discuss wood as a living system that is part of an ecosystem and subject to biological processes such as growth and decay, and reflect on how wood as a biological entity orders wood-heating practices. Such reflections, I argue, help to challenge the dominant understanding in theories of practice of materiality as consisting within a technological and human-dominated domain and may open up new ways for practice theory to contribute to discussions on alleviating climate change and reducing energy demand. This could be done, for example, by more clearly acknowledging the spatio-temporal effects of biological processes.

The chapter starts with a brief and selective discussion on how materiality has been understood in studies on energy use that draw from theories of social practice. I then link the lives of trees to the cases of wood harvesting and heating and point to biological aspects of materiality that previous accounts on wood heating and harvesting struggle to grasp. The chapter ends with a discussion on the integration of biological aspects into theories of social practice with the help of ecological anthropology and with conclusions for studies interested in steering resource use, particularly in energy use.

Human-Made Matters in Studies on Energy Use

In practice-theoretical studies on energy use, three aspects of materiality have been emphasised: the practicality of things, bodily engagements, and energy as an ingredient in practice. First, in the literature on social practices the dominant understanding of the dynamics between practices and materiality is defined through the practicality of things or flows of controlled materials. As Rinkinen et al. (2015) show, there is a strong focus on the artefactual domain of materials and how human-made objects are handled. When the object of inquiry is a practice, materials easily settle into being treated as parts of practice-as-entities or practice-as-performances. Things, however, simultaneously exist as elements in a flat and dormant 'background', as dynamic components of ongoing action, and as sites and vectors of judgement and evaluation (Rinkinen et al. 2015). Perhaps the notion of 'things' as a rather ambiguous concept has helped practice scholars to dodge the question of how practices demand different resources and different materials more generally.

Another distinctive feature of the recent studies on domestic energy use, and heating in particular, is the interest in the role of bodily and sensory engagement with heat and the active management of heat and service flows. The role of bodily and sensory engagement in both producing and demanding for energy services such as heat is discussed with concepts such as comfort, which is an affective complex of bodily sensibility with materiality (Bissell 2008). For example, heating and keeping warm are multisensory achievements and experiences, particularly because producing heat on a small scale often demands a great amount of bodily involvement. People who heat with wood frequently listen to the frost and sounds of the fire, they feel the cold and warmth in their bodies, sense the wind and moisture, smell the fire, and are surrounded by a multitude of visual things. Indeed, these bodily and affective engagements enliven people's understandings of their encounters with the more-than-human world and broaden the engagements between technologies, bodies, culture, and well-being, for instance (Wallenborn and Wilhite 2014). They further suggest that more embodied ethnographic approaches to the experience of heat and the practice of heating would be of value. From this view, involvement in heating is not an occupation but a connection, an emergent participatory openness to multiple goings-on.

Space heating can easily be seen as a practice that is centred on (human) practitioners. It is carried out to deliver services of thermal comfort to people in the shelters humans have built, in which they dwell, and which they call home. It is increasingly acknowledged, however, that the 'spread' of energy-consuming practices such as heating extends beyond: pets, animals, plants, and other material objects of care all suggest more or less distinctive 'needs' for energy services (Shove et al.

2014; Strengers et al. 2016). However, the resources used for heat provision as a material, even agential, element of practice are less often discussed in studies on energy use.

To a great extent, understanding energy use is about unpacking distinctive temporal and material engagements in everyday life. Thinking about energy demand and heating theoretically, energy is an 'ingredient' of the doing or performing of social practices (Shove and Walker 2014). Analysing the patterning of energy demand therefore involves stepping back from energy itself, as well as going beyond just an interest in the efficiencies of the technologies through which energy is converted and utilised to provide energy services (Walker 2014). This complicates the question of how materiality should be assessed and analysed. For example, in heating, energy is made usable for various services such as thermal comfort, but it also unfolds energy and heat as distinctive, manageable ingredients. Whereas practice theory emphasises heat as service provision, heating itself is also practice—not merely a physical activity but also part of broader social life. Consequently, heating can be conceptualised as a bundle of practices of managing heat and resource flows in different ways and following seasonal variations. For example, practices of logging, chopping, and stacking wood, or those of managing the wood stoves, make up the bundle of wood-heating practices (Rinkinen 2015; Rinkinen and Jalas 2016).

The above understandings of materials in practice-theoretical accounts of energy use emphasise an overly 'functional' view of things that concentrates on moments of use and utility, downplaying their uncertain, ambivalent, and contested role in the conduct of daily life and underestimating the overlapping and multiple temporal registers through which object relations are defined (Rinkinen et al. 2015). What is more, these accounts also tend to downplay the impacts and achievements of living organisms. By expanding the analysis from things in use to biological processes, and by elaborating on the idea of materiality as a process of becoming, I next emphasise the multiple, complex object-nature relations of everyday life. In what follows, I focus on the resource—wood—in mapping the intertwined becomings of wood heating.

Expanding the View on Materiality

For ecological anthropologists, human beings and other organisms are bound together in webs of life (Ingold 2005). Not until recently have some researchers started to emphasise how dynamic and even 'social' some organisms such as trees are—they are deeply embedded in networks with individuals of their own species as well as a multitude of other organisms (Wohlleben 2016). Indeed, if we look at a log, we might easily forget that, originally, this piece of wood has been a tree, a living organism with a life of its own. As illustrated next, trees perform different spatio-temporal processes and functions as part of ecosystems.

Before that discussion, as a short introduction to the practices of wood-based heating and harvesting, it is useful to mention that in Finland it is very common to use solid wood for space heating in detached houses. Even though the housing stock in Finland is relatively new, in detached houses using open fires, wood burners, and more centralised house-based systems (pellet and wood chip boilers) often forms the backbone of a system within which other technologies are integrated. Finland even witnessed a 20 per cent increase in the use of small-scale heating with wood in private houses between 1994 and 2008 (Torvelainen 2009). On a national level in 2014, wood and wood fuels were the largest source of energy overall, representing 25 per cent of total national energy consumption (Statistics Finland 2014). Small-scale wood heating is also common in other Nordic countries such as Sweden (Swedish Energy Agency 2013), and though still marginal in the UK, it is becoming more widespread there (MacLeay et al. 2014). Despite wood burning being considered a traditional way of keeping warm, new methods and technologies for heating with wood are constantly being developed. In Finland, one explanation of the popularity of wood-based heating is the high private ownership of woodlands: those people who use wood heaters often harvest the logs from their own forest or buy them from a local provider.

Harvesting, Heating, and the Lives of Trees

In this section I will point to four different aspects of the lives of trees and how they relate to the practices of wood heating and harvesting. To start with, trees grow, rest, and die. In many climates, trees do not grow continuously throughout the year but go through more active phases followed by phases of rest (Perry 1971). These phases affect logging cycles. In Finland, trees rest—or hibernate—during the coldest months, which is the preferred harvesting period. During the coldest period, the tree is not active and dries faster. This explains how the rhythms of the practice of wood management for the practice of heating are embedded in the rhythms of natural resources. Furthermore, in Finland, it is typically known that a tree is ready for harvesting when it reaches 30–80 years of age. Such a long growth cycle means that wood is not a quick fix, for example, to be used in a resource shortage, but rather demands anticipatory and multi-generational forestry planning and land use management.

Second, trees are part of ecosystems. One might say that trees carry out a number of practices as part of ecosystems: for example, they 'eat' and 'drink' (extract moisture and nutrients from the soil and sunlight), 'breathe' (through the processes of respiration and photosynthesis), and 'sunbathe' (leaves specialised for photosynthesis are arranged in such a way as to maximise their exposure to light) (see Wohlleben 2016). A number of natural conditions such as humidity, wind, sun, soil, and temperature all play an important role in shaping these practices and the lives of trees. From the point of view of the logger, this means that forest owners must pay attention to natural cycles in producing crops at an effective cost. Ecosystems such as woodlands are dynamic, and trees play an important role in reducing erosion and moderating the micro-climate. Simultaneously environmental conditions, including broader macro processes such as climate change, have an impact on what kinds of trees grow where and how fast.

Trees, like any living creature, are also exposed and vulnerable. The third aspect of the lives of trees is that they are susceptible to disease and attacks by insects and can die because of stress or other hazards such as heavy wind or a lightning strike. This means that there is a need for flexibility in practices of wood management, for example, as the resources can be ready to harvest earlier than expected and need to be adequately dried and stored. Such unpredictable affairs call for flexibility in the practices of harvesting, but also acceptance of uncertainty and the need for buffers in log stock. Trees also grow in unpredictable places and colonise areas where they may not be wanted. This requires additional thinning work, and flexibility in forest management, for example, when and where trees are cut.

Finally, as a result of many factors during their evolution as part of ecosystems, different trees develop different material qualities. Numerous species have evolved into the woody structures we know as trees: however, some are thicker and denser, taller, evergreen, or have different leaf structures and canopies. For the practice of wood heating, the different qualities of wood from these trees affect the efficiency of the burning processes, which means that local emissions can subsequently change, or there is a need for new stove designs, or the number of logs needed per fire is altered.

These examples of how trees have a life of their own show the vividness of the resource—wood—itself. They show that the ways in which wood heating has taken shape are profoundly rooted in the processes of nature. Trees as parts of local ecosystems have different spatio-temporal processes that are embedded in the bundle of wood-heating practices, as shown above, and practices develop in relation to these processes by following them or by trying to master them. In the case of wood heating and harvesting, human-made processes and the processes of biology intertwine. Thus, it's not only human aspects but also biological processes that affect these practices.

Materials of Wood Heating

In the practices of wood heating and harvesting, the biological processes of wood have a unique role. In cases of small-scale wood burning, the actual work or labour involved in making heat does not happen outside home, or in some distant, centralised factory: rather, it is typically carried out by the end user in or close to the specific locales that need to be warmed. The work involved in managing this kind of heating has a number of distinctive temporal features (Jalas and Rinkinen 2016). When organised on a small scale, wood heating requires planning and physical effort and what can be called a 'harvest-when-available and store-until-required' mentality. For instance, in Finland, people who heat with wood often source logs from their local forest or buy them from a provider who supplies them dried or wet, chopped or intact, piled or loose. Chopped logs can take two winters to 'season', and during this time they need to be stored somewhere and under conditions that enable them to dry out.

Importantly, taking a practice approach to the study of heating moves away from the unified understanding of heat as a demanded resource to its understanding as part of the temporally orchestrated set of practices in everyday life. The activities and natural processes that contribute to wood heating are dispersed across time and space and ordered by a range of temporalities from annual seasons to the daily rhythms of social life, as well as the temporalities of fuel and supply. While heating would 'ideally' allocate both warmth and work evenly across days, the work done for wood heating and the indoor temperatures provided are not evenly distributed.

Delving further into a practice approach, wood heating is made of a specific set of material elements: logs, tools, and equipment for chopping and moving wood from the forest to the stove; wood-burning technologies such as stoves, fireplaces, and central boilers; and assisting tools such as pokers. It has been argued that when we move from small-scale heating (e.g., gathering one's own logs and the use of fire stoves and open fires) to large-scale heating (centralised heat combustion and wood management), the sensitivity to ecosystems' processes and biodiversity fade (Giampietro et al. 1997). However, seeing the vitalism of the matter that is consumed does not stop us from disrupting these processes, as shown by the vast and fast degradation of woodlands (Bennett 2010). Although many aspects of wood use are tangible and controllable, it does not necessarily mean that environmental concerns are effectively integrated into practice (e.g., Petersen 2008).

The relationship between the trees, wood, the logger, and the wood heater shows the intertwinement of human and natural processes in an interesting manner. Practices of harvesting and wood heating show such threading in a distinctive and graspable way—the processes can be often seen in one human lifetime and they can be repeated over the years. Similar recurrent loops are obvious in many examples of traditional livelihoods from fishers to farmers. However, some are less easily detected as they are not noticeable within one human lifetime, and might be passed on as inter-generational knowledge.

Materiality as Becoming

On close inspection, Schatzki (2010) has already outlined how materiality in its diverse, biological form is part of social life: for him, material arrangements, which are interconnected material entities, cover physicality, composition, bio-physicality, nature, and environment. In Schatzki's (2010) view, something is material if it is physical, biological, or natural, and any thing, property, or event can be at once both social and material-natural when they are treated as social phenomena. This comes close to the idea of trees having a life of their own, and further, such an outline helps us understand the distinctively different engagements of matter in practice. Material entities have a physical composition, and biological and physical flows pass through practice-arrangement nexuses (Schatzki 2010). As Schatzki puts it, 'human sociality is constitutionally intertwined with relations to domesticated, commensal, and wild organisms' (Schatzki 2010, 129). These domesticated, commensal, and wild organisms appear differently in the theories of social practices.

Yet, it is still unclear in empirical research using theories of practice how the relationship between biological matter and processes, materials and practices can be conceptualised. Ecological anthropologists remind us that erasing the division between biological and social and seeing the biological as part of an intertwined process of becoming has already taken root (Ingold 2013). Ingold (2013) suggests observing interactions between beings and focusing on the flows that crisscross the world. For practice theorists, the challenge here is to identify the roles of different kinds of material flows within and between practices, as well as their temporal and spatial registers. The non-human world is not a passive, formable world, but it rather comprises living organisms of all kinds, which evolve. This is different from things growing—and grown by humans.

Such a stance requires different ways of knowing the open-ended processes of biosocial becoming. Materiality, in this sense, is conceptualised as a constant process of becoming as it mediates both standardisation and localisation, and transformation and rooting of practices. This requires a shift away from the individualistic idea of subject-object relations towards an understanding of things in relation to practices. Taking this view suggests that human beings do not dwell on the other side of a boundary between society and nature but in the same world that is inhabited by creatures of all kinds, human and non-human: 'all creatures, human and non-human, are fellow passengers in the one world in which they all live, and through their activities continually create the conditions for each other's existence' (Ingold 2005, 503). These ideas are shared with many theorists of practice, but suggest a more active and operative role for the non-human: non-humans continually create conditions for the existence of practices through biological processes.

Moreover, temporally sensitive accounts of material aspects of practices provide an understanding of how materialities are situated in time and how that is in part defined by the changing roles that materiality plays before, within, and beyond moments of practical action. Biological processes are vital for a living organism to live and are not necessarily always defined by 'practical' action in human terms. In this sense, natural resources as elements of practices are 'wild' despite the temporal and spatial processes that humans can observe and analyse. Forms of life, then, according to Ingold (2013, 9) are neither genetically nor culturally preconfigured but emerge as properties of dynamic self-organisation of developmental systems.

Practices are most binding when they have a dense spatial and temporal organisation—a strong teleological structure—and when the materials are actionable and in use (Schatzki 2010). Furthermore, many practices are linked to each other, but the practitioner may not consciously choose the bundle of practices he or she takes part in. Hence, assessing complex phenomena such as energy use and climate change involves looking beyond the individual to explore practices as a nexus in space and time. This calls for re-conceptualising the relation of organisms to their environments, and not thinking of them simply in terms of what they are, but what they do (Ingold 2013).

In terms of wood heating, as suggested above, looking at what wood 'does' in practices would mean not only looking at the processes of wood as a living organism—for example, how it breathes, eats, drinks, and interacts within an ecosystem—but weaving these processes into the analysis of social practices. As some might consider the processes of living organisms as practices per se (and consequently regard wood or trees as practitioners), the analysis should not end there if the aim is to better understand how social practices draw on resources. In the case of wood heating and harvesting, such extended analysis could be done by asking questions on how these biological processes are enacted in the performances and entities of such practices.

This approach complements the existing understanding of dynamic materials, for example, in Shove's (2017) account, in the sense that it does not consider resources as merely materials that are 'used up' but as something that have a life of their own and an effect (sometimes unknown) on how practices evolve. Similar to Schatzki's (2010) idea of biological and physical flows passing through practice-arrangement nexuses, the idea of practices consuming resources risks suggesting a static flow of resources 'through the practice'. Whereas Shove (2017) argues for the important task of understanding the resource-based materiality of practices, limiting these resources to only human-made ones (e.g., electricity, gas, bricks, boilers) fosters the understanding of materiality as human-made.

Conclusion

All social practices are bound with materiality in a complex and complicated manner. This chapter has considered the intersection of the social and the biological dimensions of practices by exploring the practice of wood heating in some detail. In doing so, it has raised questions about the challenges of harnessing biological ideas for thinking about practices. How do we account for the fact that something biological affects materiality and consequently also practices? How do social practice scholars analytically grasp the very mundane yet implicit processes like growth? In this chapter, I have not commented on whether trees act as carriers of practice, but rather turned attention to their biological processes and how they relate to practice. This is one step in expanding and challenging the crucial roles of materiality in practice.

The analysis of biological processes within practices might help us better understand why some practices have local characteristics and why some practices are more standardised across geographical sites. For example, given that trees are living organisms and part of ecosystems, they are subject to the conditions of local environment. It follows that the practice of wood heating will need to adapt to local conditions and consequently has its own local characteristics across sites. If this is so, are practices that draw on materials that are subject to less distinctive biological processes then more standardised? This is a notion that needs further research, but it might be of great interest to those interested in practice change. In particular, when working towards steering energy and resource use, and fostering sustainable practices, attending to the conditions of practice change is important.

Examining the practices of wood heating and harvesting, and the variety of materialities involved, it becomes obvious that the materials these practices draw on are more than a static, mouldable resource. Wood as a living system and part of ecosystems is subject to biological processes such as growth and decay, and wood as a biological entity therefore affects the practices of heating and harvesting. Against this, materiality in this chapter is conceptualised as a constant process of becoming. I have shown how the different biological spatio-temporal processes of trees as part of ecosystems relate to the bundle of wood-heating practices in a constitutive, yet under conceptualised manner. This chapter further adds to our understanding of biological materials in theories of practice by showing that biological entities mediate both standardisation and localisation, and the transformation and establishment of practices.

References

Atchison, Jenny. 2018. Thriving in the Anthropocene: Human-Weed Relations and Invasive Plant Management in Theories of Practice. In *Social Practices and More-than-Humans: Nature, Materials and Technologies*, ed. Cecily Maller and Yolande Strengers. Palgrave Macmillan.

Bennett, Jane. 2010. *Vibrant Matter. A Political Ecology of Things*. Durham, NC: Duke University Press.

Bissell, D. 2008. Comfortable Bodies: Sedentary Affects. *Environment and Planning A* 40 (7): 1697–1712.

Giampietro, M., S. Ulgiati, and D. Pimentel. 1997. Feasibility of Large-Scale Biofuel Production. *BioScience* 47 (9): 587–600.

Haraway, Donna. 2013. *Simians, Cyborgs, and Women: The Reinvention of Nature*. London: Routledge.
Ingold, Tim. 2005. Epilogue: Towards a Politics of Dwelling. *Conservation and Society 3* (2): 501.
———. 2013. Prospect. In *Biosocial Becomings: Integrating Social and Biological Anthropology*, ed. T. Ingold and G. Palsson, 1–21. Cambridge: Cambridge University Press.
Jalas, Mikko, and Jenny Rinkinen. 2016. Stacking Wood and Staying Warm: Time, Temporality and Housework Around Domestic Heating Systems. *Journal of Consumer Culture 16* (1): 43–60.
MacLeay, I., K. Harris, and A. Annut. 2014. *Digest of United Kingdom Energy Statistics 2014. A National Statistics Publication*. London: TSO. https://www.gov.uk/government/uploads/system/uploads/attachment_data/file/338750/DUKES_2014_printed.pdf.
Miller, D. 1998. Why Some Things Matter. In *Material Cultures: Why Some Things Matter*, ed. D. Miller, 3–21. London: UCL Press.
Morley, J. 2016. Technologies Within and Beyond Practices. In *The Nexus of Practices: Connections, Constellations, Practitioners*, ed. A. Hui, T. Schatzki, and E. Shove. London: Routledge.
Perry, T.O. 1971. Dormancy of Trees in Winter. *Science* 171 (3966): 29–36.
Petersen, L.K. 2008. Autonomy and Proximity in Household Heating Practices: The Case of Wood-Burning Stoves. *Journal of Environmental Policy & Planning* 10 (4): 423–438.
Rinkinen, Jenny. 2015. *Demanding Energy in Everyday Life: Insights from Wood Heating Into Theories of Social Practice*. PhD dissertation, Aalto University School of Business. Helsinki: Aalto Press.
Rinkinen, Jenny, Mikko Jalas, and Elizabeth Shove. 2015. Object Relations in Accounts of Everyday Life. *Sociology 49* (5): 870–885.
Schatzki, T. 2010. Materiality and Social Life. *Nature and Culture* 5 (2): 123–149.
Shove, E. 2017. Matters of Practice. In *The Nexus of Practices: Connections, Constellations, Practitioners*, ed. A. Hui, T. Schatzki, and E. Shove, 155–168. London: Routledge.
Shove, Elizabeth, Mika Pantzar, and Matt Watson. 2012. *The Dynamics of Social Practice: Everyday Life and How It Changes*. London: Sage Publications.
Shove, E., and G. Walker. 2014. What is Energy For? Energy Demand and Social Practice. *Theory, Culture and Society* 31 (5): 41–58.

Shove, E., G. Walker, and S. Brown. 2014. Material Culture, Room Temperature and the Social Organisation of Thermal Energy. *Journal of Material Culture 19* (2): 113–124.
Shove, E., M. Watson, and N. Spurling. 2015. Conceptualizing Connections Energy Demand, Infrastructures and Social Practices. *European Journal of Social Theory* 18 (3): 274–287.
Statistics Finland. 2014. *Energy Supply and Consumption* [e-publication]. Helsinki: Statistics Finland. Accessed August 4, 2016. http://www.stat.fi/til/ehk/2014/ehk_2014_2015-12-14_kuv_001_en.html.
Strengers, Y., L. Nicholls, and C. Maller. 2016. Curious Energy Consumers: Humans and Nonhumans in Assemblages of Household Practice. *Journal of Consumer Culture 16* (3): 761–780.
Swedish Energy Agency. 2013. *Energy Statistics for One- and Two-Dwelling Buildings in 2012*. ES 2013: 05.
Torvelainen, J. 2009. Pientalojen polttopuun käyttö 2007/2008. (The Use of Solid Wood in Detached Houses, in Finnish). *Metsätilastotiedote* 26/2009. Helsinki: Metla.
Walker, G. 2014. The Dynamics of Energy Demand: Change, Rhythm and Synchronicity. *Energy Research & Social Science* 1: 49–55.
Wallenborn, G., and H. Wilhite. 2014. Rethinking Embodied Knowledge and Household Consumption. *Energy Research & Social Science 1*: 56–64.
Warde, A. 2005. Consumption and Theories of Practice. *Journal of Consumer Culture* 5 (2): 131–153.
Wohlleben, P. 2016. *The Hidden Life of Trees: What They Feel, How They Communicate Discoveries from a Secret World*. Vancouver, BC: Greystone Books.

4

'Dynamic' Non-human Animals in Theories of Practice: Views from the Subaltern

Paula Arcari

Introduction

This chapter speaks on two key questions raised by this edited collection—first, how 'dynamic' non-humans, in this case animals, can be conceptualised in theories of social practice and, second, whether these non-humans can be conceived as performers of practices in their own right. It starts with the recognition that there is an inherently dualistic imbalance of power between human and non-human animals. Although this power may be latent, in that its effects are not always actualised, the imbalance exists as a structurally normalised and ontologically legitimised 'given' and constitutes the basis of human interactions with, and practices involving, non-human animals. For this reason, I contend that conceiving the role of non-human animals in social practices as simply 'dynamic' elides the persistently problematic dimensions of human/non-human relations. With reference to existing practice-based literature, and

P. Arcari (✉)
Centre for Urban Research, RMIT University, Melbourne, VIC, Australia
e-mail: paula.arcari@rmit.edu.au

C. Maller, Y. Strengers (eds.), *Social Practices and Dynamic Non-Humans*,
https://doi.org/10.1007/978-3-319-92189-1_4

to more critical literature that tackles questions of 'the other', I explore to what extent theories of social practices are able to address inequities between human performers and sentient 'materials' or 'objects' of practices. While this chapter is concerned with non-human animals, these sentient, objectified 'materials' can include human 'others' and may also be considered part of broader material arrangements, as in 'nature' (Schatzki 2010). For the sake of readability, I hereafter use the term 'animals' to refer to non-human animals except where I wish to emphasise the co-animality of humans.

I begin by looking at how social practice theories conceive and address power. Following this, I review practice-based studies that include animals focusing on how these animals are *present*-ed in these studies, especially in light of the previous discussion of power. In this endeavour, I explore the extent to which practice-based studies that include animals engage with more critical perspectives on human relations with human and non-human 'others'. More specifically, I ask: how are relations between human practitioners and living elements of social practices characterised? Are the ways in which effects of power may shape and direct the actions of 'others' accounted for, as Watson has argued they should (2017, 173)? Contrary to recognising and accounting for effects of power, I suggest current practice-based studies reproduce inequities between the humans and non-human worlds and contribute to their ongoing normalisation. Thinking about how social practice theories can better account for materialised 'others', I consider two routes explored in this collection: (1) conceiving non-humans as performers of practices and (2) reconceptualising their role within practices. I explain why I believe the former is neither useful nor even possible and advance the notion of the subaltern as the more effective way to acknowledge the perspective of a living 'other', who is not a practitioner, within social practices. Originally emerging from a concern with the hegemonic (and colonial) perspective of historical practices, giving voice to the subaltern 'other' can reveal inherent inequalities within practices, which then demands engagement with the systemic power relations that shape them.

Exploring this idea further, I show why this subaltern 'other' cannot be treated homogenously. Using the idea that animals constitute different kinds of 'presences' in social practices, I briefly highlight the diversity of subalternity that would need to be accounted for in a revised practice-based theory of human/animal interactions. For example, animals might be incidental or more deliberate and constitutive parts of practices. Their presence might be as a whole living animal or in de-animated parts. A consideration for these different kinds of presences, involving different relations and effects of power, would therefore provoke different kinds of questions that are important in terms of how relations between practitioners and materialised others are problematised—questions concerning who is materialised, how, and what about them is materialised, in other words how they are being ontologised.

In conclusion, I reason that social practice theories do not provide an accurate or useful way of conceptualising the activities of animals and other living non-humans. Although recognising that non-humans have practices is a valid observation, theorising them in this way contributes to the elision of a fundamentally anthropocentric world view. Instead of directing efforts towards conceiving non-humans as practitioners, I show how a consideration for the subaltern might inject practice-based approaches with more critical capacity, specifically in terms of being able to confront an inherent human-centrism and address associated effects of power.

Power in Practices

Up until recently, the social practice literature has not specifically engaged with theories of power. Elsewhere, power is commonly theorised hierarchically with a focus on groups or individuals who 'hold' more or less power. This perspective is incompatible with a focus on practices popularly conceived as non-hierarchical—adhering rather to a flat ontology (Schatzki 2016). However, more Foucauldian theorisations of power as a dispersed, indeterminate phenomenon of social relations that is 'employed and exercised through a net-like organisation' (Foucault 1980, 98) have found resonance with Schatzki (2005, 2015) and Watson (2017).

Power is implicit in certain theorisation of practices prior to these works by Schatzki and Watson. For example, it is implicit in Schatzki's earlier assertion that 'social order' is an outcome of practices (1996) and in the 'codes' and 'patterns of conduct' that, through 'social steering' (Halkier 2004, 29), are conceived as shaping social norms (Warde 2005, 141; Barnett et al. 2005, 27). Identifying an affinity with Foucault in his approach to power, and prefiguring Watson's statement about power shaping or directing other's actions (2017, 173), Schatzki describes 'how actions structure *others*' possibilities' (2005, 479, emphasis added). He continues: 'the notion of power thus captures the responsibility that actions bear for the differential access that other people have to the possibilities carried in practice-arrangement bundles' (479). Thus, in common with critical formulations of power and oppression (Foucault 1980; Young 1990; Deckha 2008), Schatzki does not conceive these actions as the result of individual, deliberate choices. He describes power as an 'omnipresent feature of human coexistence' (2015, 20), where large social phenomena are 'anchored' by 'power centres', which link, and affect, multiple sites of action (2015, 18). This sense of power permeating social practices through systemically normalised inequalities is reflected also in Walker's (2013) effort to locate justice in social practices, where he notes, 'inequality and injustice have to do with variety and with the differential patterns of goods and bads (in a broad sense) that are reproduced in the ways that lives are lived' (2013, 183) or in the 'churn of innumerable moments of practices' (Watson 2017, 10).

However, power remains generalised, unlocated, and vaguely articulated in social practice theories. The foregoing considerations of power stop short of naming the sorts of 'broader structures' and 'large social phenomena', such as patriarchy, 'racism', and other systemic forms of discrimination, by which unequal relations are encoded and reproduced in practices. Consequently, they do not venture beyond loose observations of power as, for example, an 'omnipresent' 'effect of collective activity' (Watson 2017, 180) with 'consequences' in terms of 'issues of access and inclusion/exclusion' (Walker 2013, 181) to ask *how* these consequences come about and what exactly is the nature of these issues of access, inclusion, and exclusion. A more nuanced interrogation of the nature of power, which explores how it is constituted, would enable social practice theories to consider all the ways in which 'others' might be involved in, recruited into, or affected by a practice. Without such questions, social practice theories risk inadvertently reproducing the dualisms they seek to collapse.

Although I am here referring specifically to the human/nature and human/animal dualisms, they are all interconnected and co-constitutive. As Kim observes, power is multidimensional (2015, 11).

It is only a small step from conceptions of social order and of conduct being shaped, including intimations of limitation, constraint, and exclusion, to theorisations of domination and oppression. Indeed, some form of domination and oppression is implicit in the effect of power, as Watson implies: 'purposive change will involve engaging in and with existing *dominant power relations*' (2017, 10, emphasis added). Refining and honing articulations of power would provide an opportunity for practice theories to productively engage with the range of 'social locations' (Deckha 2008) already well-recognised as sources of inequality, including 'race', gender, sexuality, class, age, ability, migration status, religion, and, I would add, species. The perspectives of those whose role in social practices is shaped by these locations—who are materialised as part of practices (e.g. as labour or entertainment) or who might be considered 'externalities' of practice (i.e. subject to the indirect consequences of practices)—need to be considered.

It is these alternate narratives that I contend are not accounted for in the practice literature. Hence, how these 'others' come to be constituted as materials or externalities of practices is not questioned. Unless these *a priori* constitutions are viewed critically, studies of social practices risk reflecting, and further normalising, unequal relations shaped by systemic privileges—whether social, cultural, economic, religious, or species. This is not to say that other social and cultural theories are always critically reflexive and attentive to the reproduction of socially constituted terms and concepts. However, social practice theories have made, and are making, a novel and invaluable contribution to understandings of the social world and to tackling some of its pervasive dualisms. To strengthen this contribution, and uphold claims that they are actively collapsing dualisms, these theories need to attend to those dualisms that remain largely unchanged. Sharpening their critical capacity regarding the materialisation of living 'others', human and non-human, offers one way to further enhance the value of these approaches and the insights derived from them.

Accepting that power is a 'feature of human coexistence' (Schatzki 2015, 20), how, then, are relations between human and non-human, specifically animals, conceived in studies of social practices? How are animals *present*-ed? Are they merely biotic materials, resources, and/or entertainment in human-centred endeavours? Are the conditions of their materialisation, and its consequences in terms of shaping and directing their actions, acknowledged? These questions are explored in the next section.

Animals in Social Practices

In most cases, the term non-human is used in practice-based approaches as a catch-all for both material objects and living organisms, including animals. In much of this literature, the non-human is loosely conceived as 'entities', 'assets', 'artefacts', or 'resources' (Cetina 2001; Reckwitz 2002; Domaneschi 2012; Walker et al. 2014). Sometimes the term implicitly excludes animals (and presumably other living entities) as in Poel and Bakker's conception of practices as always including 'both human bodies and nonhuman objects or materials' (2016, 133).[1] In their view, practice competency thus refers to 'the use or dealing with material objects by more or less skilled human beings' (133). The non-human may also reference a generalised 'nature', 'the environment', 'living organisms' (Gherardi 2017; Røpke 2009), or Reckwitz's 'nonhuman creatures' (2002, 207). Ambiguity is the rule rather than the exception.

Two recent edited volumes on practice theory (Spaargaren et al. 2016; Hui et al. 2017) are notable for not explicitly addressing non-human plant or animal life, despite the non-human being named as a key theme in the former and the latter encompassing 'posthumanist' theory. The only exception is Morley's reference to animals as one among many non-human energy sources (2017, 85)—i.e. a human resource. Although the volume by Hui et al. includes reference to 'organisms', 'microorganisms', and 'nature', the non-human being recognised seems primarily to be

[1] Previous to this, these authors note that whether practices might exist in the animal world poses an interesting question for discussion. However, the potential for animals to be the non-human objects and materials of practices is not entertained.

non-living. Only occasionally in practice-based studies are more specific living non-humans referred to, such as plants (Hitchings and Jones 2004), bacteria (Cetina 2001), microorganisms (Maller 2017), fish (Law and Lien 2013; Roe 2006), pets and 'pests' (Strengers et al. 2016), and 'bodies that are eaten' (Roe 2006, 104).

In some of these studies, the authors are explicitly challenging anthropocentric conceptions of practices (Hitchings and Jones 2004) and practitioners (Strengers et al. 2016). However, across all studies it is the pre-determined function of these non-humans in social practices—their socio-materiality—that is foregrounded in a way that does not critically question the construction of this function (e.g. as food, pet, or pest). This treatment is also common in ANT, materiality, and posthuman approaches. This is not to negate the value of this literature in revealing interesting nuances in relations between a human 'owner' and a 'companion' animal (Haraway 2008); farmers and 'their livestock', 'dairy cows', and 'beef cattle' (Yarwood and Evans 2000; Law and Miele 2011); humans and salmon in fish-farming practices (Law and Lien 2013); homeowners, 'pets', and 'pests' in energy consumption practices (Strengers et al. 2016); or horse-breeding practices and their associated 'thoroughbreds' and 'broodmares' (Schatzki 2010). However, the ontological operation that takes place to determine the roles of these animals in social practices remains mostly unquestioned.

In his exploration of horse-breeding practices, Schatzki (2010, 141) is uncritical of the systemic power relations and human-centrism that normalise conceptions of animals based on their uses. He observes 'how material entities are intelligible to humans is tied to the practices the latter carry on'. Noting that these arrangements 'might also be meaningful to certain non-humans that are elements of them (e.g. horses)', he then cedes that 'this controversial issue cannot be presently addressed' (Schatzki 2010, 141). Law and Miele similarly emphasise that '*practices enact animals*', that 'animals are *an effect of practices*' (2011, 59, emphases in original). But this is stated as the basis from which the authors undertake a non-critical exploration of the enactment of different versions of 'cow', not as a process deserving of further unpacking.

What this brief overview illustrates is that social practice theories uncritically and unproblematically treat all kinds of animal 'stuff' more or less equally, whether living or dead. The constitution of animals as a 'backdrop'—a material resource base—for human activities is unquestioningly accepted. The studies reviewed here reflect and reproduce a human-centric view of animal existence and thereby reinforce the normalisation of distinctly power-based relations. In the same way, a certain practitioner-centrism in theories of social practice could be conceived as reproducing and further normalising imbalances of power with human 'others' who are positioned as materials of practices.

While practice-based approaches have been rightly acknowledged for bringing nature and its 'active, relational materiality' (Goodman 2001, 195) back into the social sciences, and 'allow[ing] us to recognize animals as 'subject'' (Taylor 2011, 210), their potential has only been applied in a limited fashion—one that overlooks the consequences to 'others' of their materialisation in certain practices and how this happens. As Pick (2012), referencing Bennett, observes, recognising the 'vitality' of matter does not by itself presume an enhanced sense of multispecies responsibility (2012, 77), just as recognising the 'dynamism' of non-humans does not automatically challenge normalised dualisms and systemic effects of power. This oversight has significant and tangible implications for the lives of sentient others by continually re-inscribing socially constituted boundaries and limits to their existence. Consequently, with reference to Watson (2017), to the extent that social practice theories are not presently accounting for the ways in which the lives of animals (and human 'others') are shaped and directed, they can be accused of a degree of ethical anaemia.

Perhaps, then, it is time for practice-based perspectives to take up this 'controversial' issue, as Schatzki (2010) frames it, by considering how the intelligibility of living 'others' within practices is constituted and what alternate meanings these others might attach to the practices of which they are part. This is, in effect, the same reorientation in academic thinking that the recognition of human-centrism is demanding more broadly. Human-centrism is probably a largely inescapable and fundamental feature of research. However, some approaches, like social practice theories, are already oriented towards critiquing normalised dualisms and are therefore better placed than others to recognise and ameliorate its effects. Naturally, this is especially important when the practices in question involve sentient others.

In the case of animal others, it requires a degree of imagination to consider the alternate, non-materialised meanings they might attach to themselves and the practices into which they are integrated with little to no possibility of refusal. However, using what is understood about power—how its effects shape and direct the lives of 'others'—and having almost shrugged off Cartesian contestations regarding animals' capacities to suffer and experience emotions (Barron and Klein 2016), it seems increasingly negligent *not* to summon this imagination. It would, therefore, be a productive move for social practice theories to more fully confront effects of power in social practices by allowing materialised 'others' to be viewed as more than their function, as something theoretically other than material objects or background schemas. It is perhaps partly as a consequence of this theoretical framing (of materials, arrangements, and infrastructures) that the instrumentalising of 'others' has been normalised in social practice theories, or at least perpetuated.

How, then, can social practice theories better account for dynamic non-human (and human) others? It is to this question that I turn in the next two sections.

Animals *with* Practices or Animals *Within* Practices?

This volume suggests two avenues for better accounting for non-humans in theories of social practice. The first is to conceive them as performers of social practices, and the second involves improving understandings, and theoretical conceptions, of their performative role within practices or 'in the enactment of social practices' (Maller and Strengers 2018). It is not suggested that these are mutually exclusive. However, in this section, I explain why I do not think the former offers much value either empirically or theoretically. This is especially the case if the primary goal is more accurately, and ethically, representing living non-human 'others' as opposed to broadening the applicability of social practice theories. This provides the basis for my contention that living others are more usefully accounted for by focusing on their role *within* social practices and drawing on the notion of the subaltern to articulate that in a more nuanced way than is presently available.

In theories of social practice, the notion that non-humans can be practitioners of practices is occasionally alluded to in passing and without further enquiry as to what that actually means. For example, in *The Nexus of Practices* (Hui et al. 2017), Alkemeyer and Buschmann (2017, 15) speak of participants of practices as both human and non-human, implying a theoretical equality, while Nicolini, in the same volume, refers to human and non-human carriers of practices (2017, 105). There is no further information about what or who these non-humans might be. Strengers et al. (2016, 12) more specifically regard household pets as performers of practices, and other authors in this collection also explore the question of whether non-humans can perform practices (e.g. Kuijer, Strengers, Gram-Hanssen). In light of my previous observation that animals' lives are constituted by, and in relation to, human-defined social practices, I briefly examine in what way non-human animals could be described as practitioners, carriers, or performers of practices.

If animals are to be thought as practice performers, how can we humans go about defining these practices? From their perspective, humans are an assemblage of material elements and meanings—from our physical and technological presence to the trace of our smell—that shape their everyday lives. In *Being a Beast*, Charles Foster notes the poverty of human language when it comes to describing the nuances of the sensory worlds in which animals excel. He narrates the story of a man who placed his palm on a badger path for one minute. Some 11 hours later, a male (boar) and a female (sow) badger come along the path. The boar sniffs and makes a detour around the scent while the sow stops and goes no further. Having lived underground as a badger for several weeks, Foster arrives at the closest he can to a non-human interpretation of how that palm differently shaped the practices of these two badgers:

> Along the path there was a wall, built of scent particles sticking to the veins of dead leaves and the squashed casts of long-dead worms. To the boar, that wall had definite dimensions: he could skirt round the edge and go into the world beyond. For the sow, made conservative and fearful by maternal responsibility, the wall was indefinitely high and long and the world beyond it unthinkable. (2016, 59)

Though he describes how the trace of human scent *appeared* to affect each badger, Foster cannot know the meanings and understandings that shaped their different responses or what other materials or associated practices might have constituted them. Neither can he ask the badgers. He can only infer. Importantly though, he does not need to know, and his narrative is almost stronger for this unknowability of the badgers' thoughts of the 'world beyond'—a final refuge from human inquisitiveness/acquisitiveness. Why, then, seek to apply a theory that requires such conjecture—attempting to mine the purpose and intent behind animals' lives in order to articulate their practices? I see this as qualitatively different to considering the alternate perspective animals would have on their materialised roles *within* practices—a consideration that would challenge normalised designations of food, pest, invasive species, and so on, and acknowledge they are likely to hold a different view of their ontological purpose. I therefore suggest, rather contentiously, that efforts to better understand the minds of other species in this way, however well-intentioned, are ultimately human-centric. We are at a point—identified as the sixth extinction (Kolbert 2014)—where we should perhaps consider the possibility that animals would fare better if we relinquished the desire to know and understand them better, which itself can be read as another aspect of the unlimited accessibility of their lives to which humans feel entitled. We can observe when they appear well, sick, happy, afraid, anxious, and so on and reflect on how social practices contribute to their individual and collective well-being. Anything more, for me, raises alerts as to the human intentions behind it.

Aside from being impossible (in terms of ascertaining meanings and understandings), I therefore see questionable value in trying to define animals and other living non-humans as practitioners in their own practices. This is not just because it would be based on human-centric conjecture and ventriloquising of the 'other' (Armstrong 2002; Jazeel 2014), but because it prompts a question regarding the purpose of such an exercise. Typically, practice-based studies are aimed at recognising conditions/opportunities for change or some kind of intervention. But in studies that attempt to describe and intervene in 'non-human practices', where the 'practitioner/performer' cannot speak, we can ask who really the practitioner is. In contrast to studies where the voices of human practitioners are present, such studies run a greater risk of becoming practices in which humans are the practitioners and animals, once again, the materials. I do not dismiss the value of taking a less human-centric approach to the animals and other living entities we share the earth with; however, I believe there are other ways to approach this theoretically that do not require humans conjuring non-human meanings and understandings of the world and which acknowledge the *social* practices of which these intentions are part. In sum, animals and other non-humans may shape the social practices which they, incidentally or deliberately, find themselves part of as living materials, but they cannot be theorised as having their own practices. The human perspective will always impinge on conceptions of non-human practices, and indeed, attempts to conceive their lives from a practice perspective cannot be separated from the social practices they are thus being made part of. To reference Jonathan Safran Foer (2009, 46), 'it is [inevitably] to touch upon how we understand what it means to be us and not them. It is to ask, 'What is a human?''.

While rejecting the value of conceiving of 'non-human practices', I believe social practice theories need to more fully acknowledge a prevailing human-centrism and better account for effects of power that shape practices, and are in turn reproduced by them. Framing the role of animals (and other human and non-human 'others') in social practices as subaltern provides one way of foregrounding for critical analysis their constitution as materials. This refocus would address the social/species locations, dualisms, and other mechanisms and effects of power that support these constitutions, and which contribute to ongoing inequities between human and human/non-human others.

The Subaltern

Theorisations of the subaltern offer a way of thinking about how 'others', in this case animals, and their alternate perspectives, might be recognised and accounted for in social practice theories. By acknowledging the subaltern as distinct from other non-human materials, the actual or theoretical voices of these 'others' may be included, making them more truly dynamic, at least from a less human-centric perspective.

Subalternity is understood as 'a condition of subordination brought about by colonization or other forms of economic, social, racial, linguistic, and/or cultural dominance' (Beverley 1999, Frontmatter). Species may equally be included as a basis for subordination. The term has already been applied by critical animal scholars to explore the transnational pet-cloning industry (Yoon 2016), and the 'post-colonial animal' (Armstrong 2002)—where Armstrong recognises a 'common antagonist' (the imperialising human ego) in postcolonial and animal studies. Narayanan (2017) develops the notion of 'subaltern animism' to explore how street canines in Indian cities claim and occupy space. Law and Lien (2013, 11) also refer to 'subaltern beasts', described as those 'might-have-beens and nearlies', or 'not-quite realised orders of animals', which have shifted *outside* human-animal practices. For Law and Lien, the term subaltern denotes those animals that no longer serve any human function, being no longer 'knowable to the human' (11). Animals used as part of social practices are therefore not subaltern, but rather fully 'realised'. This represents a twisting of the term, whereby instead of giving voice and visibility to the perspective of 'others', they are cemented further in their subalternity by depicting it as the 'proper' order of things. The marginalising effect of power relations on others, through the control and acquisition of their space, time, physicality, and biology, is specifically what the term subaltern is intended to draw attention to (Jazeel 2014).

In her study of veganism as a posthuman ethic, Giraud (2013) refers to the traditional binaries that posthumanism seeks to challenge—human/animal, culture/nature, subject/object—and describes what is normatively regarded as the subordinate of each pairing as the subaltern term. She suggests these oppositions could be challenged 'by practices that both focus ethical attention on the subaltern term in these pairings and challenge the positioning of this term as subaltern' (53).[2] Introducing this notion of the subaltern to practice theories, which claim a similar anti-dualistic intent, could therefore draw attention to and question the constitution of living materials (animals, nature, objects, and 'other'

[2] Although Giraud does not refer to social practices, she does refer to practices of veganism and meat consumption and also theoretical practices and the (humanist) epistemologies that legitimate them. It is this second meaning of practices, relating to theoretical approaches, that I assume she is referring to here.

humans) and thus problematise the practices of which they are part. This is something that has heretofore not been considered in theorisations of social practices.

Acknowledging the subaltern allows these 'others' to be rendered more visible. Currently, they are largely invisible, or not recognised as distinct from materials, and may have limited access to practices as a result of unequal power but also limited power within or as a result of practices. This conceptualisation still allows for animal bodies, and parts thereof, to be considered material elements of practices, but with the added recognition of the 'absent referents' (Adams 2015 [1990])—the living subaltern animal. More nuanced interpretations of relations between practice elements, and the constitution of practices, may result from this re-conceptualisation of practices.

There are many practices that are shaped by and reproduce effects of power (such as domination and oppression) in which subaltern 'others' are implicated without consent and used as material resources based on their attributed function in the practice, for instance, practices associated with possession or control of others such as human trafficking, slavery, child labour, and the sex trade. Power, and its twin aspects of domination and oppression, is especially prominent in such practices, as 'the act of possessing opens up the prospect of also being possessed' (Howitt 2002, 300). Whether understood as the 'subaltern' element of practices, or by another means, *how* these 'others' come to be constituted as 'other' is a key component in their subsequent objectification and materialisation, by which their animate and inanimate bodies are made available as material elements across a range of practices. Theorisations of oppression—which encompass (among others) colonialism, critical race studies, feminism, and notions of patriarchy—attend to the 'how' of 'how power is done' (Watson 2017, 177). By allowing living elements to be differently recognised and 'heard', these theories could make a constructive contribution to articulations of power within social practice theories. This would enable practice theorists to more directly engage with questions of inequality and social justice, confronting issues of 'race', socio-economic status, 'gender', sexual orientation, ability, religion, and also species.

Admitting the presence, and voice, of the subaltern in certain practice-based studies, in place of what (or who) would have previously been conceived in terms of materiality, allows for the recognition that there are everyday social practices that are (for many) exploitative and harmful. In addition, it admits the possibility of alternate and even dissenting perspectives on more commonly studied practices where the presence of a subaltern status may be harder to perceive, for example, practices relating to pet keeping (DeGrazia 2002), community gardening (Guthman 2008), waste disposal (Pellow 2004), firefighting (Klein 2017), and building design (Moore et al. 2017).

The subaltern as a concept is not without critique. The act of identifying a group of humans or non-humans in terms of their subordination to a dominantly held view of the world is not only nominal but performative. In being described as subaltern and without voice (Spivak 1988), their location and treatment as such may be reinforced, ignoring all the ways in which, aside from perhaps not even recognising the practices in which they are designated subaltern, they may regard themselves as actively resisting or exerting some other kind of agency of their own (as per Narayanan 2017). Armstrong notes that just as the actions and voices of non-Western peoples may be underestimated by such fatalism, 'animal studies have demonstrated that agency in human-animal interactions proves complex and irrepressible and cannot be reduced to the hollow phantasm that Berger and Baudrillard see in the figure of the pet, the zoo animal, the stuffed toy and the Disney character' (2002, 416). This does not negate the value of the subaltern as a theoretical device, but rather highlights the need for a nuanced and non-homogenous conceptualisation of these materialised 'others' in practice-based studies—one that allows their voices to be heard while recognising their diversity.

Considering subalternity in terms of the type of 'presence' allocated to a living 'other' (human or non-human), as part of a social practice, provides one way of thinking about how these diverse voices might be included. For instance, what is it about the 'other' that is being materialised, used, or affected in some way? Is it their aliveness (e.g. animals used for hunting, entertainment, labour, education), their designation as 'other' (e.g. as nature, pest, non-native), or their biology (e.g. animals used for scientific/medical research, fashion, 'food', and in myriad everyday products)[3] that is materialised? Does the materialisation apply to their whole body or disaggregated parts? To what extent is their life shaped and directed as a result? Is this shaping and directing incidental (e.g. a result of land use changes) or more deliberate and constitutive of the practice (e.g. 'meat' production)? Extending how practices involving living, materialised 'others'—human or non-human—are theorised in this way is, I believe, essential. Without critically questioning how they are constituted as materials and how this affects the conditions of their (and others') existence, especially when that materialisation is also commodified, social practice theories risk a certain ethical passivity. Practices can tend to be described and explored as accepted, albeit changing, aspects of everyday life, further normalising fundamentally unequal and problematic relations between practitioners and (once) living elements. The proposed Anthropocene lends this point added significance with its associated questions concerning humans' historically normalised, dualistic approach to nature and non-humans—an approach that social practice theories already seek to contest. The concept of the subaltern therefore provides a way in which practice-based research could directly address effects of power that are reproduced and reinforced in social practices.

Bringing the idea of the subaltern into studies of social practices might begin with a series of questions something like:

- Does the practice in question shape or direct the lives of living others—that is, 'who's' rather than 'what's'?
- Are these lives affected incidentally—as externalities of practices, or deliberately—as materials that are more and less central to the constitution of the practice?

[3] For example, products that include leather, feathers, wool, silk, or fur; cosmetics, vitamins, medicines, and pharmaceuticals; anything that uses glue or contains bone, ivory, horn, shell, or other animal parts; and tennis rackets, bone china, plywood, perfume, tyres, paint brushes, and even computers are just some of the everyday objects and products that include the traces of animals used in their manufacture or testing.

- What social conditions support and legitimate the materialisation of these others?
- How might these others describe their constitution in relation to the practice in question? Would they see themselves as something other than materials? Or perhaps as practitioners in different practices?[4]
- How might 'other' understandings and meanings of a practice be incorporated into analyses to mitigate their potential practitioner- and human-centredness, and guard against the reproduction of structurally normalised power relations?

For example, with reference to Poel and Bakker's (2016) exploration of the role of material elements in the constitution of tennis as a practice, how different might the tone of their analysis have been with an acknowledgement of the subaltern, absent*ed* animals that some of these materials (viz. rackets and balls) rely on. This is not to imply that all practice-based enquiries ought to account for how each material is constituted, but it does raise a question regarding scope. To what extent should each 'material' be accepted as already 'done', thereby reproducing and further normalising any inherent power relations, and to what extent should its constitution be problematised? This will largely depend on the primary purpose and focus of the study. However, in some circumstances, greater acknowledgement of the (human and non-human) lives that may have contributed to certain materials could help guard against the aforementioned ethical passivity. The broader complex of practices that rely to greater and lesser degrees on the materialisation of animals (both living and dead) is vast.

The addition of the subaltern is, I argue, compatible with existing theorisations of social practices. Unlike attempts to conceive of non-humans as performers of practices, reconceptualising their materiality within social practices as subaltern can go some way to acknowledging the effects of power in practices. It also offers some defence against the uncritical reproduction of these effects.

---

[4] For example, would Atchison's (2018) invasive species describe their own practices as invasive?

Conclusion

The impulse for this chapter derived from my observation that most animals have no choice in whether or not to be part of social practices. As Palmer (2001, 344) comments, large numbers of animals are not 'free to opt for other possibilities'. They are involved in relationships with humans that are 'persistent and nonreversible' (344) and that no amount of stratagems or tricks can reverse (355). This is ultimately true for any animal drawn incidentally or deliberately into the world of human-oriented social practices, whether as a consequence of their animated aliveness or their biology. It is also true for any human (or hybrid) 'other' similarly drawn into practices within which their freedom to opt for another possibility is curtailed. As part of these practices, these 'others' do not enjoy the same freedoms as human 'practitioners' and their status is therefore not equal. The extent to which non-humans can be regarded as dynamic in social practices that are conceived and constituted around human purposes is therefore *a priori* ontologically limited—more limited for some than for others. However, in the case of animals, there is a problem of circular reasoning in expecting these animals to occupy anything other than secondary status in analyses of practices perceived and defined from a wholly human perspective.

Based on this, and in light of the structurally normalised and ontologically legitimised imbalance of power between human and non-human animals, I have questioned the way that animals are currently conceived and *present*-ed, or absented, in studies of social practices. By first examining how the issue of power is addressed in the practice literature, and then exploring how this literature accounts for, or *presents*, non-humans and particularly animals, I identified an unavoidable human-centrism in social practice theories. This observation led me to question how these theories might better account for non-humans. My response is to suggest the notion of the subaltern as a potentially useful theoretical adjunct to practice-based analyses—one that can address the problem of human-centrism while avoiding the flawed and ultimately misleading route of conceiving non-humans as performers/carriers of practices.

The subaltern, however, should not be treated as a homogenous or fixed category of 'other'. By identifying different kinds of animal presence, from incidental to deliberate, and in whole or disaggregated parts, I showed how the materiality of animals is constituted in subtly different ways and with different ontological implications. This suggests an ethical responsibility to 'listen' to the (human or non-human) subaltern in different ways, depending on the kind of presence they have in practices, whether incidental or deliberate, and the extent to which their actions are shaped and directed.

Introducing the notion of a subaltern presence allows the theoretical status of non-human (and human) 'others' as materials to be maintained and critically foregrounded. This in turn allows the subaltern perspective to be accounted for while avoiding any potential obfuscation of the prevailing human-centrism of social practices. Human-centrism is intended as a descriptive rather than derogatory term, meaning that social practice theories can be nothing other than human-centric. Attempting to make them less so by relating non-human practices to our own actually confounds their human-centrism. Embracing it, and speaking to it directly through the subaltern, is, I argue, the more effective way to confront prevailing dualisms and their co-constitutive effects of power. More important therefore than determining whether animals have practices is re-visiting social practice theories as they are currently conceived so that they allow for, and preferably encourage, a critical stance towards constitutions of practitioners and materials. Specifically, I contend that when the material of a practice is a 'who' (human or non-human), then there is an ethical responsibility to pay greater attention to how that 'who' is so constituted and how they are positioned in relation to the 'practitioner' of the practice. That responsibility is perhaps easier to recognise when those 'who's' are human, but further conceptual work is required to acknowledge the alternate non-human perspectives subsumed within social practice theories that cannot be accounted for in the same way. This work can begin with a more critical orientation towards the constitution of animal elements and materials that rely on them—an orientation that the questions posed in the previous section are intended to instigate.

In sum, recognising that social practice theories are intrinsically and unavoidably human-centric, and that 'materials' of practices are not created as such but are socially constituted, is the first step. Next, when those materials are human or non-human 'who's', researchers ought to at least be sensitive to the conditions of their constitution as 'materials'—for example, social/species locations, persistent dualisms, and effects of power. These locations, dualisms, and effects of power are complicit in today's global-scale environmental problems, such as biodiversity loss, mass extinction, and climate change, and social problems such as mal- and undernutrition, poverty, and discrimination. If they hope to make any meaningful posthuman contribution to dismantling all dualisms, practice-based studies should at least consider whether it is ethical to accept materialisations of certain humans and non-humans, perhaps thereby reproducing normalised effects of power, or whether it is incumbent on the analysis to contribute to challenging (or at least not reproducing) problematic conceptions of the 'other'. This will depend on the nature and purpose of the study, but researchers should nevertheless be mindful of how the everyday reproduction of socially normalised locations and dualisms can shore up their normalisation and reinforce associated effects of power. By granting a space where the presences and voices of materialised subaltern 'others' can be acknowledged, social practice theories gain greater critical capacity to recognise and challenge effects of power and thereby steer a path away from ethical passivity.

References

Adams, Carol. 2015 [1990]. *The Sexual Politics of Meat*. New York; London; Oxford; New Delhi; Sydney: Bloomsbury Academic.

Alkemeyer, Thomas, and Nikolaus Buschmann. 2017. Learning In and Across Practices: Enablement as Subjectivation. In *The Nexus of Practices: Connections, Constellations, Practitioners*, ed. Allison Hui, Theodore Schatzki, and Elizabeth Shove, 8–23. London: Routledge.

Armstrong, Philip. 2002. The Postcolonial Animal. *Society and Animals* 10 (4): 413–419.

Atchison, Jennifer. 2018. Thriving in the Anthropocene: Understanding Human-Weed Relations and Invasive Plant Management Using Theories of Practice. In *Social Practices and Dynamic Non-humans: Nature, Materials and Technologies*, ed. Cecily Maller and Yolande Strengers, 25–46. Cham, Switzerland: Palgrave Macmillan.

Barnett, Clive, Paul Cloke, Nick Clarke, and Alice Malpass. 2005. Consuming Ethics: Articulating the Subjects and Spaces of Ethical Consumption. *Antipode* 37 (1): 23–45.
Barron, Andrew B., and Colin Klein. 2016. What Insects can Tell Us about the Origins of Consciousness. *Proceedings of the National Academy of Sciences* 113 (18): 4900–4908.
Beverley, John. 1999. *Subalternity and Representation: Arguments in Cultural Theory*. Durham, NC and London: Duke University Press.
Cetina, Karin Knorr. 2001. Objectual Practice. In *The Practice Turn in Contemporary Theory*, ed. Theodore Schatzki, Karin Knorr Certina, and Elke von Savigny, 184–197. London: Routledge.
Deckha, Maneesha. 2008. Intersectionality and Posthumanist Visions of Equality. *Wisconsin Journal of Law, Gender and Society* Spring (1): 249–267.
DeGrazia, David. 2002. *Animal Rights: A Very Short Introduction*. Oxford: Oxford University Press.
Domaneschi, Lorenzo. 2012. Food Social Practices: Theory of Practice and the New Battlefield of Food Quality. *Journal of Consumer Culture* 12 (3): 306–322.
Foer, Jonathan Safran. 2009. *Eating Animals*. London: Penguin Books.
Foster, Charles. 2016. *Being a Beast: An Intimate and Radical Look at Nature*. London: Profile Books.
Foucault, Michel. 1980. *Power/Knowledge: Selected Interviews and Other Writings, 1972–1977*. New York: Pantheon Books.
Gherardi, Silvia. 2017. Sociomateriality in Posthuman Practice Theory. In *The Nexus of Practices: Connections, Constellations, Practitioners*, ed. Allison Hui, Theodore Schatzki, and Elizabeth Shove, 38–51. London: Routledge.
Giraud, Eva. 2013. Veganism as Affirmative Biopolitics: Moving Towards a Posthumanist Ethics? *PhaenEx* 8 (2): 47–79.
Goodman, David. 2001. Ontology Matters: The Relational Materiality of Nature and Agro-Food Studies. *Sociologia Ruralis* 41 (2): 182–200.
Guthman, Julie. 2008. Bringing Good Food to Others: Investigating the Subjects of Alternative Food Practice. *Cultural Geographies* 15 (4): 431–447.
Halkier, Bente. 2004. Handling Food-related Risks: Political Agency and Governmentality. In *The Politics of Food*, ed. Marianne E. Lien and Brigitte Nerlich, 21–38. Oxford: Berg.
Haraway, Donna J. 2008. *When Species Meet*. Minneapolis, MN: University of Minnesota Press.
Hitchings, Russell, and Verity Jones. 2004. Living with Plants and the Exploration of Botanical Encounter Within Human Geographic Research Practice. *Ethics, Place & Environment* 7 (1-2): 3–18.

Howitt, Richard. 2002. Scale and the Other: Levinas and Geography. *Geoforum* 33 (3): 299–313.

Hui, Allison, Theodore Schatzki, and Elizabeth Shove. 2017. *The Nexus of Practices: Connections, Constellations, Practitioners*. London: Routledge.

Jazeel, Tariq. 2014. Subaltern Geographies: Geographical Knowledge and Postcolonial Strategy. *Singapore Journal of Tropical Geography* 35 (1): 88–103.

Kim, Claire Jean. 2015. *Dangerous Crossings: Race, Species and Nature in a Multicultural Age*. New York: Cambridge University Press.

Klein, Naomi. 2017. *No Is Not Enough: Resisting Trump's Shock Politics and Winning the World We Need*. Chicago: Haymarket Books.

Kolbert, Elizabeth. 2014. *The Sixth Extinction: An Unnatural History*. New York: Henry Holt & Co.

Law, John, and Marianne Lien. 2013. Slippery: Field Notes on Empirical Ontology. *Social Studies of Science* 43 (3): 363–378.

Law, John, and Mara Miele. 2011. Animal Practices. In *Human and Other Animals: Critical Perspectives*, ed. Bob Carter and Nickie Charles, 50–65. Basingstoke: Palgrave Macmillan.

Maller, Cecily. 2017. Epigenetics, Theories of Social Practices and Lifestyle Disease. In *The Nexus of Practices: Connections, Constellations, Practitioners*, ed. Allison Hui, Theodore Schatzki, and Elizabeth Shove, 68–80. London: Routledge.

Maller, Cecily, and Yolande Strengers. 2018. Dynamic Non-humans in a Changing World. In *Social Practices and Dynamic Non-humans: Nature, Materials and Technologies*, ed. Cecily Maller and Yolande Strengers, 1–21. Cham, Switzerland: Palgrave Macmillan.

Moore, Trivess, Ian Ridley, Yolande Strengers, Cecily Maller, and Ralph Horne. 2017. Dwelling Performance and Adaptive Summer Comfort in Low-income Australian Households. *Building Research & Information* 45 (4): 443–456.

Morley, Janine. 2017. Technologies Within and Beyond Practices. In *The Nexus of Practices: Connections, Constellations, Practitioners*, ed. Allison Hui, Theodore Schatzki, and Elizabeth Shove, 81–97. London: Routledge.

Narayanan, Yamini. 2017. Street Dogs at the Intersection of Colonialism and Informality: 'Subaltern Animism' as a Posthuman Critique of Indian Cities. *Environment and Planning D: Society and Space* 35 (3): 475–494.

Nicolini, Davide. 2017. Is Small the Only Beautiful? Making Sense of 'Large Phenomena' from a Practice-based Perspective. In *The Nexus of Practices: Connections, Constellations, Practitioners*, ed. Allison Hui, Theodore Schatzki, and Elizabeth Shove, 98–113. London: Routledge.

Palmer, Clare. 2001. 'Taming the Wild Profusion of Existing Things'? A Study of Foucault, Power, and Human/Animal Relationships. *Environmental Communication: A Journal of Nature and Culture* 23: 399–358.
Pellow, David Naguib. 2004. *Garbage Wars: The Struggle for Environmental Justice in Chicago*. Cambridge, MA and London: MIT Press.
Pick, Anat. 2012. Turning to Animals Between Love and Law. *New Formations* 76: 68–86.
Poel, Hugo van der, and Sven Bakker. 2016. Grounding the Practice: Material Elements in the Constitution of Tennis Practices. In *Practice Theory and Research: Exploring the Dynamics of Social Life*, ed. Gert Spaargaren, Don Weenink, and Machiel Lamers, 131–150. London: Routledge.
Reckwitz, Andreas. 2002. The Status of the "Material" in Theories of Culture: From "Social Structure" to 'Artefacts'. *Journal for the Theory of Social Behaviour* 32 (2): 195–217.
Roe, Emma J. 2006. Things Becoming Food and the Embodied, Material Practices of an Organic Food Consumer. *Sociologia Ruralis* 46 (2): 104–121.
Røpke, Inge. 2009. Theories of Practice—New Inspiration for Ecological Economic Studies on Consumption. *Ecological Economics* 68 (10): 2490–2497.
Schatzki, Theodore R. 1996. *Social Practices: A Wittgensteinian Approach to Human Activity and the Social.* Cambridge, UK and New York: Cambridge University Press.
———. 2005. Peripheral Vision: The Sites of Organizations. *Organization Studies* 26 (3): 465–484.
Schatzki, Theodore. 2010. Materiality and Social Life. *Nature and Culture* 5 (2): 123–149.
———. 2015. Practices, Governance and Sustainability. In *Social practices, Interventions and Sustainability: Beyond Behaviour Change*, ed. Yolande Strengers and Cecily Maller, 15–30. London: Routledge.
———. 2016. Practice Theory as Flat Ontology. In *Practice Theory and Research: Exploring the Dynamics of Social Life*, ed. Gert Spaargaren, Don Weenink, and Machiel Lamers, 28–42. London: Routledge.
Spaargaren, Gert, Don Weenink, and Machiel Lamers. 2016. *Practice Theory and Research: Exploring the Dynamics of Social Life*. London: Routledge.
Spivak, Gayatri Chakravorty. 1988. Can the Subaltern Speak? In *Marxism and the Interpretation of Culture*, ed. Cary Nelson and Lawrence Grossberg, 271–313. Basingstoke: Macmillan Education.

Strengers, Yolande, Larissa Nicholls, and Cecily Maller. 2016. Curious Energy Consumers: Humans and Nonhumans in Assemblages of Household Practice. *Journal of Consumer Culture* 16 (3): 761–780.

Taylor, Nik. 2011. Can Sociology Contribute to the Emancipation of Animals? In *Theorizing Animals: Rethinking Humanimal Relations*, ed. Nik Taylor and Tania Signal, 203–220. Leiden and Boston: Brill.

Walker, Gordon. 2013. Inequality, Sustainability and Capability: Locating Justice in Social Practice. In *Sustainable Practices: Social Theory and Climate Change*, ed. Elizabeth Shove and Nicola Spurling, 181–196. London: Routledge.

Walker, Gordon, Elizabeth Shove, and Sam Brown. 2014. How Does Air Conditioning Become 'Needed'? A Case Study of Routes, Rationales and Dynamics. *Energy Research & Social Science* 4: 1–9.

Warde, Alan. 2005. Consumption and Theories of Practice. *Journal of Consumer Culture* 5 (2): 131–153.

Watson, Matt. 2017. Placing Power in Practice Theory. In *The Nexus of Practices: Connections, Constellations and Practitioners*, ed. Allison Hui, Elizabeth Shove, and Theodore Schatzki, 169–182. London: Routledge.

Yarwood, Richard, and Nick Evans. 2000. Taking Stock of Farm Animals and Rurality. In *Animal Spaces, Beastly Places: Critical Geographies*, ed. Chris Philo and Chris Wilbert, 99–115. London: Routledge.

Yoon, Hyaesin. 2016. Disappearing Bitches: Canine Affect and Postcolonial Bioethics. *Configurations* 24: 351–374.

Young, Iris Marion. 1990. *Justice and the Politics of Difference*. Princeton, NJ: Princeton University Press.

5

Dynamic Bodies in Theories of Social Practice: Vibrant Materials and More-than-Human Assemblages

Cecily Maller

Introduction

This chapter is concerned with how bodies are represented in theories of social practice. It begins from two main points of contention: first, that many aspects of bodies that make them dynamic, vibrant, and different appear under-acknowledged in work using theories of practice and, second, that social practice theories are generally concerned with a narrow definition of bodies, where human bodies are prioritised in the performance,[1] trajectories, and persistence of practices (aside from this collection). There are many ways to discuss bodies and theories of social practice that surpass the capacity of one book chapter. My interest here concerns taking a more detailed look at the materialities and agencies of

[1] In discussing performance in this chapter, I refer to the doing, or carrying out, of practices. Although I do not address it, I acknowledge there is an important difference between performance and performativity, as argued by Butler (1990).

C. Maller (✉)
Centre for Urban Research, RMIT University, Melbourne, VIC, Australia
e-mail: cecily.maller@rmit.edu.au

C. Maller, Y. Strengers (eds.), *Social Practices and Dynamic Non-Humans*,
https://doi.org/10.1007/978-3-319-92189-1_5

bodies, how bodies are conceptualised in theories of practice, and how other theoretical resources can be drawn on to revitalise or rethink multiple dynamic bodies in practice performances, entities, and the making of social worlds. In doing so, I turn to the 'more-than-human' thinking of new materialisms, assemblage theories, and science and technology studies (STS), to begin to unpack what bodies do in practices, and what practices do to bodies. Stemming from a growing interest in ethics and justice in urban contexts, I am interested in working towards a deeper understanding of the outcomes and effects of practices on bodies, both human and non-human.[2]

As highlighted in Chapter 1, one starting point for this collection is that the materiality and agencies of dynamic non-humans may not be sufficiently accounted for in their usual classifications as 'objects', 'materials', or 'material arrangements' in theories of social practice. Recent practice theories are post-humanist[3] in their focus on relations between bodies, things, and materials (e.g., Reckwitz 2002b; Schatzki 2002; Shove et al. 2012). However, they present particular views of materiality that may be limiting regarding certain lines of enquiry about bodies, their dynamism and materialities, and how these are represented in practices. As Shove (2017) observes, material elements in theories of practice can be given the same or equal status, regardless of how directly they are implicated in the carrying out of practices. She explains further that "there is rather less analysis of the range and variety of material relationships involved" (Shove 2017, 156). Although this may not always be a problem, it matters when we want to know why and how specific patterns of practice arise (Shove 2017). The potential constraints of not distinguishing between material elements and different types of materiality become important when considering the visceral impacts of practices (or the outcomes of practices) on bodies, both human and non-human, for example, in dealing with widespread, complex health issues like obesity (Maller 2017), questions of more-than-human ethics (Ginn 2014) and animal rights (see Chapter 4), or social and environmental justice (Walker 2015).

[2] My interest is constrained to living rather than non-living bodies.

[3] Post-humanism has many definitions. I consider post-humanism to be thinking and theories that move beyond human subjectivity and human centrism, as explained by Schatzki (2010).

In their analysis of how bodies have been conceptualised in energy consumption studies, rather than seeing bodies as 'natural' or 'fixed entities', Wallenborn and Wilhite (2014, 57) conceptualise bodies as "at the intersection of the cultural and natural, the human and non-human, the individual and collective". Similarly, as Reckwitz (2002a, 251) reminds us, "routinized bodily performances are the site of the social". This is my entry point for this chapter. Through the arguments presented in the sections outlined below, the chapter shows how, along with other elements or ingredients, practices are comprised of vibrant, dynamic multiple bodies and bodily materials. I am not arguing that bodies should become the central focus of practice scholarship, rather that there may be ways to extend theories of practice when it comes to understanding how bodies and practices are co-constituted. Such a move might be useful for investigating certain topics or questions, for example, in rethinking bodies, can theories of practice answer politically inspired questions about which bodies come to matter and why (e.g., Butler (1993))? The next section introduces my two contentions regarding how bodies are dealt with in theories of practice, before providing a brief outline of the remainder of the chapter.

Two Contentions About Bodies

The first point I contend is that the dynamic, vital, and varying characteristics of human bodies tend to be underplayed in work drawing on theories of practice. Generally dynamism and change in theories of social practice are attributed to practices—or practice bundles, complexes, or nexuses—rather than materials or bodies (e.g., Shove et al. 2012). This position may be explained by theories of practice moving beyond focusing on what individual people (or bodies) do, to instead analyse practices and their configurations (Shove et al. 2012). It may also be because questions about bodies and 'biology' have traditionally been considered outside of the realm of interest to social theorists. As Schatzki (2010) acknowledges, organisms and our relations to them may not be of interest in questions about social phenomena (although I argue this is changing; see Meloni (2014) and as I discuss further below).

Bodies are accounted for in most practice theories through notions of performance and carriage of practices (Shove et al. 2012; Reckwitz 2002a) and where 'doings' and 'sayings' (Schatzki 2002, 23) and 'competences' and 'meanings' (Shove et al. 2012) are considered fundamental components or elements of practice entities. As Reckwitz (2002a) clarifies, social practices require *both* bodily and mental activities. Although materials and materiality are one defining characteristic in current practice literature (e.g., Blue and Spurling 2017; Hitchings and Lee 2008; Rinkinen et al. 2015; Schatzki 2010; Shove et al. 2012; Strengers and Maller 2012), bodies are not always included in definitions of materiality. For example, in Shove and colleagues' (Shove et al. 2012) description of practices, there are three key practice elements: materials, competences, and meanings. In defining the material element of practices, they write, it includes the "things, technologies, tangible physical entities, and the stuff of which objects are made" (Shove et al. 2012, 14). Bodies may be included by "tangible physical entities", but in this instance, Shove and colleagues do not specify whether this is the case.

Schatzki (2010) has a different way of theorising materials and bodies in his formulations of social practices. Rather than including a material dimension *within* practices, Schatzki (2010) describes social worlds as comprising nexuses of practices and material arrangements, linked together over time-spaces and as co-constitutive of each other. Material arrangements are "a set of interconnected material entities" made up of four different types: "humans, artifacts, organisms, and things of nature" (Schatzki 2010, 129). Schatzki (2010, 137) thereby overtly includes human and non-human bodies, acknowledging that the materiality and composition "of human bodies and of [other living and non-living] entities" are important determinants, or preconfigurations, of practices. He goes on to explain further that "organisms and matter-energy flow through practice-arrangement nexuses"[4] and that non-human organisms, their materiality and their sociality, are part of society (Schatzki 2010, 138). However, in subsuming bodies as part of the mesh of material arrangements, or by focusing on practices as the key site of action and dynamism, has some of the vibrancy and agency of bodies been lost or

[4] See Oppermann and Walker this collection, Chapter 7.

discounted in theories of practice? Along these lines, I have argued elsewhere how human bodies have a 'present-absent' status in practice theories (Maller 2017). They are clearly present as the 'carriers' of practices, essential for the continuing performance and persistence of practice entities; however, bodies are absent when their qualities and differences are overlooked (Maller 2017). Without intending to resort to objectification, what I mean is that the materiality, physicality, composition, health status, age, gender, race, ability, and other intersecting aspects of bodies are rarely foregrounded in work drawing on theories of practice. This may be due to the sorts of questions asked to date; however, here I argue there is room for further consideration of bodies.

In the social practice literature, it is generally agreed that practices and bodies[5] are co-constitutive, in that practices require certain things from bodies and, at the same time, are the result of bodies being trained, and performing, in particular ways (Wallenborn and Wilhite 2014; Reckwitz 2002a; Schatzki 2010; Mol 2002; Wainwright and Turner 2006). One conclusion from this is that bodies are made by practices. For example, in repeated practice performances, bodies are changed as they clearly become skilled in carrying out practices, contributing to the persistence and continuity of those practices (assuming all elements and other conditions for continued 'survival' of the practice entity are met). But as time passes, the bodies of practitioners age, are modified, or deteriorate, which potentially reduce their capacity to carry out and perform certain practices, for instance, doing professional sports such as ballet (Wainwright and Turner 2006). If new bodies are not recruited to practices, then, in the language of practice theories, due to a shortage of carriers, those practices will need to adapt or risk "fossilisation" (Shove and Pantzar 2005), "disappearance", or "decay" (Shove et al. 2012). Hence, just as bodies are made by practices, so are practices made by bodies.

Conceiving of bodies changing over time encourages further reflection on the capacities of different bodies to perform practices, how bodies are made different through practices, and what this line of thinking might imply for efforts to change practice trajectories for social and environmental arguments or to address questions of ethics, equality, and justice.

[5] 'Bodies' therefore refers to body/minds (Reckwitz 2002a).

For example, bodies at different life stages and differently abled, gendered, classed, or racialised bodies[6] will not only perform practices differently but are also likely to perform different practices due to (lack of) opportunity and the structuring of effects of practice bundles and complexes (Shove et al. 2012). For example, people on low incomes may not be able to afford to participate in practices such as digital photography or floorball (e.g., Shove and Pantzar 2007). This means that practice trajectories are impacted through the lives, abilities, and capacities of practitioners or carriers over time and space, just as their lives are impacted by the practices they do, or do not, participate in. In differentiating between different bodies, their materialities, and agencies, theories of practice could have more to contribute to research addressing complex questions of health inequality and related social issues.

My second, related, contention is that although practice theories look beyond human bodies to consider relations and interactions between bodies, things, and material arrangements, they generally remain concerned with the performance and competencies of human bodies in practices across time and space rather than acknowledging bodies of other kinds (e.g., Reckwitz 2002b; Schatzki 2002; Shove et al. 2012) (for exceptions, see Strengers et al. (2016) and Schatzki (2010), discussed below). There are many other bodies that are potentially recruited to, implicated in, or are at least arguably an indispensable part of, practices. For example, gardening and dog-walking practices directly involve the bodies of non-human species (including plants, soil microbiota, pollinators, and dogs respectively), without whom these practices would not be able to be carried out. In the majority of work using practice theory, non-humans tend to be classed as materials or material arrangements, with their agencies, dynamism, rights, and vitalities potentially homogenised or unnoticed. The emphasis on human bodies in the performance and carriage of practices means that other bodies may not be counted or valued. In many cases, depending on the questions or topic of interest, the way bodies are currently accounted for may be sufficient in theories of practice. However, as Mol (2002, viii) argues, what is at stake is ontological politics, meaning "the way in which problems are framed, bodies are

[6] For discussions of the concept of intersectionality, see Crenshaw (1991).

shaped, the way that lives are pushed and pulled into one shape or another" matters. Different bodies do, and are made by, different practices; this applies not just to human bodies but also to non-human ones.

Taking ontological politics into account (see Chapter 6), and inspired by the feminist scholarship of Mol (2002) and Butler (1993), some questions that this discussion raises in regard to theories of social practice and human and non-human bodies include: *how* are bodies made through practices? Which bodies are implicated in what practices? What do bodies do in, and how are they used by, practices? How are bodily differences accounted for, or accommodated, in practices? When and why does this matter? More practical or empirical questions include: what do human and non-human relations constituted through practices mean for human health and wellbeing? And what do these same relations mean for non-human ethics, rights, and welfare? Although I do not directly address them here, consideration of these and other questions could help theories of practice rediscover, or be reoriented towards, (human and non-human) bodies and resolve potential homogenising and normalising tendencies where all bodies may be presumed to be equal.

To expand on and address the two contentions outlined here, in the remainder of the chapter, I draw on threads from several different sources of literature, including new materialisms and assemblage thinking. In doing so, I provide two interrelated ways bodies could be given more attention in social practice scholarship (thereby allowing that what makes sense for different lines of enquiry may vary). One approach is to conceive bodies as vibrant and dynamic materials. The other is to conceive them as more-than-human assemblages. Although the sources of literature that each of these sections draws on are extensive, there is too little scope in a single chapter to discuss the strengths, weaknesses, and complexities of each. The next section considers what bodies can bring to practices through their materiality and physicality. Using ideas from new materialisms, political ecology, and health, I discuss how human bodies provide different forms of vital materials in and to practices, for example, in practices of eating and sleeping. Turning to assemblage thinking, the final section discusses how most bodies, including humans, are part of complex more-than-human assemblages. It also considers how these ideas might be usefully brought into conversation with theories of social practice to recognise other forms of non-human bodily agency and dynamism, and expand the reach of topics and issues explored using these theories.

Bodies as Vibrant and Dynamic Materials in Practices

The material contributions bodies provide to practices are regular, continuous, and often imperceptible (Bennett 2010); in many ways they form a background hum of internal activity compared to some of the more visible aspects of practice performances, such as 'doings and sayings'. Nonetheless, these contributions are essential to practices that depend on living, breathing practitioners for manifestation in the world, as well as the usual material and other requisite elements of practices. In some cases, the material contributions of bodies are more clearly evident in practices such as pregnancy and childbirth, where parents, particularly mothers, contribute a large range of materials and non-humans to a developing foetus, including genetic information, immune and defence systems, proteins, and microbiota.

More than a straightforward, transactional contribution however, the materiality provided by bodies to practices is "vibrant, vital, energetic, lively, quivering, vibratory, evanescent, and effluescent" (Bennett 2010, 112).[7] This vibrancy manifests itself in several ways in social practices. For example, it is evident in the transformation of matter ingested as food in eating practices or the water and air that are processed in, and used by, bodies to stay alive (Bennett 2010); through growth, change, and ageing as bodies are physically sculpted by multiple, repeated practice performances over time, enhancing or diminishing competences and skills and impacting health and wellbeing (Maller 2017); through practice memories inscribed on bodies materially and mentally for future reference and resurrection (Maller and Strengers 2015); and through intergenerational

[7] Bennett (2010) argues that all matter, whether living or dead, is vibrant after philosophers Spinoza and Bergson.

material effects that are passed on to future generations of practitioners through reproductive and epigenetic processes (Maller 2017). Bennett (2010, 50) summarises the liveliness of this activity nicely using the example of metabolisation:

> [where] the outside and inside mingle and recombine, renders more plausible the idea of a vital materiality. It reveals the swarm of activity subsisting below and within formed bodies and recalcitrant things, a vitality obscured by our conceptual habit of dividing the world into inorganic matter and organic life.

When viewed in this way, it is hard to return to bodies conceptualised just as carriers or practitioners of practices. Bodies develop skills, grow, age, reproduce, and die. In other words they not only can shape, but are shaped by, practices in intense reciprocal arrangements as pointed out above.

The discussion so far has hinted at a perceived intersection or boundary between the social and biological sciences by considering how some bodily processes and activities are essential to practice performances and social life more generally. Such 'biological' processes are generally considered to be well outside of the domain of social practices, as much as biology and sociology are separate disciplines. However, new materialist and other scholars have been calling for the dissolution of boundaries between the 'natural' and 'social' sciences for some time (Barad 2007; Coole and Frost 2010b; Pickering 1993; Meloni 2014). More than ever disciplinary boundaries need to be crossed or transcended so that practitioners and scholars can find and develop innovative thinking and solutions for complex global problems, such as climate change and obesity. It is increasingly recognised that what were once considered separate worlds, or domains, are more frequently coming into contact where 'unprecedented things' (Coole and Frost 2010a, 4) are happening to bodies and the world, such as stem cell and genetic interventions in human health, the rise of artificial intelligence and automation, and pandemics caused by bacteria crossing species divides. In this vein, but in a more everyday context, the following paragraphs explore two practices that involve both biology and sociality: sleeping and eating.

Sleep is something that most animals partake in and humans are no different. As well as eating, drinking, sex, and other bodily processes and activities that have become routinised through practices, sleep is a biological imperative and an activity essential to maintain health. Whether it is a practice or not, however, has been subject to deliberation due to its primarily recognised biological basis and function, evidenced by the large volume of work on sleep in the literature, the bulk of which is based in medical and natural science (Taylor 1993). There is also considerable research on sleep disorders due to the serious effects of sleep deprivation at all life stages and on many aspects of health, wellbeing, and social responsibilities and relationships (e.g., Chokroverty 2017).

The exact functions of sleep still remain a mystery despite decades of research; however, there are a number of theories about what sleep is for and what it does. These include body and brain tissue restoration, energy conservation, memory reinforcement and consolidation, the maintenance of synaptic and neuronal network integrity, immune and endocrinal regulation, thermoregulation, and the removal or clearing of metabolic waste (Chokroverty 2017). It is also widely known that without sufficient sleep, bodies are impaired in performance, memory, and immune suppression among other deleterious effects (Chokroverty 2017). Although even to a keen observer sleeping bodies do not seem very active or dynamic, there are clearly a number of key internal processes that happen during sleep that are vital for keeping bodies functioning and active during waking hours and for replenishing energy reserves and restoring mental capacities, all of which are necessary for bodies to carry and perform practices.

Despite sleep having multiple important biological functions, it is also socially defined (Oskar and O'Connor 2005; Taylor 1993). There are a number of ways sleep becomes social and subsequently is considered a social practice. At the outset, there is a difference between being asleep and 'doing sleeping' (Taylor 1993, 464). Being asleep means body-minds go into a semi- or unconscious state (Chokroverty 2017) and are seemingly at rest. In this state, all other practices and activities are suspended as 'being asleep' is not compatible with the performance of other practices. In other words, in the state of sleep, bodies are undertaking "the business of what sleep is actually for" (Taylor 1993, 464). 'Doing sleeping' however is much more social. Doing sleeping, including how long it is done for, is defined by and means different things according to culture, age/life stage, health status, and income (Taylor 1993). For example, the sleep of babies and patients is often an observed and monitored activity, while the sleep of healthy adults is not. As Taylor (1993, 464) writes, sleep is a "culturally-defined and influenced practice" with specific meanings that create a "discourse of dormancy". So pervasive is this discourse that some meanings of sleep have crossed over to be associated with the meanings of other practices such as sex and death, for example, 'sleeping around' (Taylor 1993). Questions of how, when, where, and with whom bodies sleep should therefore be of central interest to those investigating the nocturnal aspects of social life (Taylor 1993). As well as meanings, certain skills are also required in the doing of sleeping. Babies and children must learn to sleep, as do adults recovering from sleep disorders (Chokroverty 2017). Lastly, the material elements of sleep, although varied, at a minimum usually involve a room or place to sleep, something to sleep on, covers or bedding, and certain clothing. So rather than being an 'extra-sociological experience' (Taylor 1993, 464) confined to the realms of biology or medicine, sleeping is very much a social practice associated with a vast number of biological doings.

Eating is another example of a biologically based practice that involves dynamic bodily processes and liveliness. Several authors have already argued convincingly that eating is a social practice (e.g., Bourdieu 1984; Delormier et al. 2009; Halkier and Jensen 2011; Warde 2014) so I will not dwell on the different elements that comprise eating practices, nor on the other practices it is connected to, such as shopping and cooking. Instead, I focus on the dynamic interactions between bodies, foodstuffs, and practices closely associated with the ingestion and conversion of food to show how intertwined social practices can be with biology and bodily processes.

Like sleep, most animals must eat in order to stay healthy and alive. Eating essentially involves the ingestion and mastication of edible matter that is transformed by bodies into energy and used in processes to grow, reproduce, and maintain muscles, bones, and tissues and produce hormones and so on; "eating becomes a series of mutual transformations in which the border between inside and outside becomes blurred" (Bennett 2007, 134). There are many ways eating involves dynamic bodily functions. First, eating requires the activities and actions of various parts and organs to break down, digest, and process food, from the mouth, stomach, intestines to the rectum and anus. Second, once broken down or 'liberated' into constituent parts such as monosaccharides and amino acids, food materials are modified for transportation via blood or lymphatic circulations and taken to different parts of the body where they are taken up or utilised by cells and tissues (Medeiros and Wildman 2011). Third, parts of food that are indigestible or not required are processed as waste, and fourth, any matter that is not required to carry out practices and other activities is transformed into 'excess' body fat/weight. As Bennett writes (2010, 44) "when bodies and food interact and perform eating, digestion, waste elimination and a whole raft of other metabolic and biological/physiological processes [take place]". These sophisticated doings are clearly biological, metabolic, biochemical, and dynamic, involving many different parts of bodies and bodily systems, but they are also inherently social in how they manifest themselves in social practices such as commensality, practices of feeding of others such as children and babies, and waste treatment and toileting. Like doing sleeping, doing eating is also essential in maintaining bodies so that other practices can be performed. Likewise, the meanings and 'how' of doing eating are defined through practices. For example, what is considered polite to do in the case of bodies burping or passing wind, and what these bodily functions mean, is social.

In these examples of sleeping and eating, bodies are made through practices that require, and contribute, both biological and social components. Hence rather than classifying some things bodies do as being largely biological and others as social, more subtle, fluid, dynamic, and potentially useful insights are revealed when the activities bodies do are considered together as bio-social doings and as part of practices. The final section of this chapter takes a slightly different but related view of bodies in social practices by considering them as more-than-human assemblages.

Bodies as More-than-Human Assemblages

The previous section argued that bodies are vibrant materials in, and contribute materials and biological processes to, practices in dynamic ways. This section discusses a related way the dynamism of bodies can be considered in theories of practice and offers a means of rethinking bodies without problematically recentring human individuals. Always focusing on humans as the carriers of practices can discount the roles of non-human bodies and materialities, and maintains boundaries between biological and social thought, as well as dualisms, such as between humans and non-humans. In human geography and related disciplines, the idea of 'more-than-human' has become popular to overcome boundaries between humans and non-humans (Abram 1997; Whatmore 2002). In essence, more-than-human thinking and theories argue for studying, understanding, and experiencing the world in less human-centred and more material, performative, relational, or dynamic ways (Maller 2018). Hence this section makes the case that bodies can be conceptualised as more-than-human assemblages (broadly interpreted) in the performance and ongoing trajectories of practice entities. This move is not such a stretch for theories of practice; as Shove (2017, 157) observes in her discussion of the different roles materials play in practices, objects, things, and materials are not isolated entities but are "always integrated within and always inseparable from more extensive assemblages".

Assemblage thinking was first written about by Deleuze and Guattari (1987). In general, ideas of assemblage are used to focus on processes rather than end products or states, such as the relations and distributed agencies between multiple entities, emphasising emergence, multiplicity, and indeterminacy (Anderson and McFarlane 2011; DeLanda 2006). One useful outcome of thinking in 'assemblages' is that boundaries are immediately called into question, not only boundaries between organisms or disciplines but between bodily parts and their components. This type of conceptualisation unsettles the illusion perpetuated in some research that human bodies or their parts have fixed or discrete edges in the world, unconnected to life, matter, and action around them. As well as the material contributions human bodies make to the doing of practices (as described in the previous section), they also play host to vast numbers of microorganisms and use many things, objects, technologies, or other materials in the performance of practices. As Mol (2002, viii) writes, "bodies are an intricately coordinated crowd". This section works through more-than-human thinking using the example of microbiomes to illustrate how human bodies are never just one body, and furthermore, that bodies are always connected in intricate ways to various types of materials and matter. Therefore, I consider that human and non-human bodies carry and perform social practices as more-than-human assemblages.

Authors as diverse as Yong (2016), Bennett (2010), Mol (2002), and Haraway (2008) have argued that humans are never just one body. In their article published in *Nature*, Turnbaugh et al. (2007, 804) observe "the microorganisms that live inside and on humans (known as the microbiota) are estimated to outnumber human somatic and germ cells by a factor of ten". These large communities that live in and on human bodies consist of a variety of microorganisms including eukaryotes, bacteria, and viruses. A microbiome is the aggregate collection of microorganisms living in association with the human body. The profundity of their sheer numbers is worth contemplating in regard to the materiality and dynamism of bodies (human and non-human) in social practices. These microbes are generally not harmful to humans; conversely they are essential for maintaining health and staying alive (Yong 2016). For example, they produce vitamins that humans do not have the genes to make, break down food in the gut to extract nutrients needed for survival, and teach immune systems how to recognise dangerous invaders (Yong 2016). Microbiota also regulate drug metabolism, interact with immune systems, and may even influence human actions such as risk taking, psychomotor skills, and driving practices (Flegr 2007).

Returning to practices of eating, the nutrient value of food to human bodies as it is processed in the digestive tract is affected by the microbiota present leading to varying health and wellbeing outcomes (Turnbaugh et al. 2007). For example, microbiota species present in human digestive systems influence whether certain diseases, obesity or malnutrition, occur (also dependent on the type and amount of food eaten). This means there are constant, dynamic interactions amongst assemblages of human bodies and their microbiota that shape the physical manifestation and health status of humans. Thinking in assemblages can be applied at more scales than the micro ones described here. For example, Wallenborn (2013) has described similar ideas in his notion of the 'extended body', which includes human bodies and everything else required to perform a practice. In relation to sustainability and energy-consuming practices, bodies may be extended "through different 'prosthesis'" in their use of things like mobile phones, appliances, and computers (Wallenborn and Wilhite 2014, 61). Similar examples of 'extended bodies' or assemblages involving living non-humans mentioned earlier were dog-walking and gardening.

So what does the expansive thinking of assemblage mean for theories of social practice? One thing it can do is distribute agency more evenly among humans, non-humans, and things to better account for the non-human dynamism and materiality that comprises, and is comprised by, practices. Another outcome is to recognise bodies other than humans and their vital role in some practices. Studying practices that directly involve the bodies of humans and non-human species could reveal new insights about how bodies are implicated in practices and lead to different understandings of performance (e.g., the idea of co-performance as explored in Chapter 10). As the examples in the present chapter have shown, human bodies contain, and are made of, many bodies of non-humans and are connected to a range of materials and matter. Recognising that microbiota and biological processes such as digestion play a role in certain practices expands the type of action that can potentially be studied (or subject to intervention by health professionals and policy makers), and helps to demonstrate that in practice performances, "a multiplicity of scales" is always present (Thrift 2008, 17) where "things in the background are of necessity tied to things in the foreground and to the ongoing mobilisation of things in action" (Shove 2017, 158). Such thinking might help find new ways to understand complex health problems such as obesity, where there are intense, ongoing bio-social doings at play in the manifestation of the condition that to date has failed to be solved by conventional approaches. To sum up, assemblage thinking is useful in foregrounding the dynamism of human and non-human bodies and materialities and leads to consideration of the idea that practices are not only carried and performed by humans but are carried and performed by more-than-human bodily assemblages.

Discussion and Conclusion

This chapter began with two main arguments about how bodies are conceptualised in theories of social practice: that the dynamism, vibrancy, and differences of bodies appear under-acknowledged and that social practice theories prioritise or foreground human bodies over others. Thinking of bodies as vibrant materials and more-than-human assemblages, or more overtly acknowledging their dynamism, materialities, and agencies in practices, helps to foreground and distinguish different bodies, and bodies in general, from other materialities.

Theories of practice already have some of the conceptual resources required to do this. For example, Shove (2017) has argued for similar distinctions to be made in related discussions about different non-living materialities implicated in energy demand and how the same things can be used differently within and between practices. As she writes, thinking more flexibly or fluidly about materials in practices "makes it possible to show how things switch between roles and to recognise that things which have a background or infrastructural relation to certain practices may be more directly engaged in the conduct of others" (Shove 2017, 164). Relatedly, Schatzki's (2010) concept of material arrangements already differentiates between different bodies in that it includes humans, organisms, and 'things of nature' (although note the possible bio-social divide whereby humans are neither organisms nor part of nature). However, I have argued in this chapter that there is scope to expand how bodies are conceptualised in theories of practice by drawing on the 'more-than-human' ideas of new materialisms and assemblage theories, as well as looking to other disciplines to understand what different types of bodies and materialities might mean for practices and what practices mean for bodies.

Although in many cases the existing treatment of bodies by social practice theories may be sufficient, it can become a sticking point when it comes to understanding issues of equity, justice, health, and how human and non-human bodies are impacted by practices. Consideration of questions related to these and other social phenomena could open up new territory for theories of practice to explore. Further consideration of the biological or bio-social underpinnings and impacts of some practices could also help reveal clues about how practices might change, where their trajectories might go, or how they could be changed or improved for social and environmental reasons. Along these lines, I provided two interrelated ways bodies could become more present in social practice theories, without necessarily centring them. These were bodies as vibrant, dynamic materials and bodies as more-than-human assemblages. The value of these conceptualisations is that they allow for closer examination and understanding of the materiality and constant dynamism of bodies in practices and the impact of practices on human and non-human bodies over time and space. Some non-humans are clearly more than just passive background to practices; rather they are active participants and agents in bio-social worlds.

I end this chapter on a provocation: if it is accepted that practices are performed by bodily assemblages of humans and non-humans as has been suggested here, is it not too great a leap to consider that in some circumstances, social practices might also be performed by non-humans? The present volume seeks to tackle this provocation head-on, such that ideas posited by contributors about co-performance by non-humans and outright non-human performance are being experimented with (see Chapters 2, 4, 10, 11, 12). Such notions pose further exciting possibilities for rethinking how we understand social practices and how policy makers and researchers could think about attempting to change practices for positive social and environmental change.

Acknowledgements These ideas were first explored during the symposium 'Animals, Automated Devices and Ecosystems' held at RMIT Europe in October 2015. Further refinement has occurred writing them up here, and I sincerely thank Gay Hawkins, Stanley Blue, and Yolande Strengers for their incisive and insightful suggestions for improvements and for being constructively critical.

References

Abram, David. 1997. *The Spell of the Sensuous: Perception and Language in a More-than-Human World*. New York: Vintage Books.

Anderson, Ben, and Colin McFarlane. 2011. Assemblage and Geography. *Area* 43 (2): 124–127. https://doi.org/10.1111/j.1475-4762.2011.01004.x.

Barad, Karen. 2007. *Meeting the Universe Halfway: Quantum Physics and the Entanglement of Matter and Meaning*. Durham, NC: Duke University Press.

Bennett, Jane. 2007. Edible Matter. *New Left Review* 45 (May–June): 133–145.

———. 2010. *Vibrant Matter: A Political Ecology of Things*. Durham, NC: Duke University Press.

Blue, Stanley, and Nicola Spurling. 2017. Qualities of Connective Tissue in Hospital Life: How Complexes of Practice Change. In *The Nexus of Practices: Connections, Constellations, Practitioners*, ed. Allison Hui, Theodore R. Schatzki, and Elizabeth Shove, 24–37. Abingdon, Oxon: Routledge.

Bourdieu, Pierre. 1984. *Distinction: A Social Critique of the Judgement of Taste*. Cambridge, MA: Harvard University Press.

Butler, J. 1990. *Gender Trouble: Feminism and the Subversion of Identity*. New York: Routledge.

Butler, Judith. 1993. *Bodies that Matter: On the Discursive Limits of "Sex"*. New York: Routledge.

Chokroverty, Sudhansu, ed. 2017. *Sleep Disorders Medicine: Basic Science, Technical Considerations and Clinical Aspects*. New York: Springer.

Coole, Diana, and Samantha Frost. 2010a. Introducing the New Materialisms. In *New Materialisms: Ontology, Agency, and Politics*, ed. D. Coole, S. Frost, and J. Bennett, 1–43. Durham, NC: Duke University Press.

———. 2010b. *New Materialisms: Ontology, Agency, and Politics*. Durham, NC: Duke University Press.

Crenshaw, Kimberle. 1991. Mapping the Margins: Intersectionality, Identity Politics, and Violence against Women of Color. *Stanford Law Review* 43 (6): 1241–1299. https://doi.org/10.2307/1229039.

DeLanda, Manuel. 2006. *A New Philosophy of Society: Assemblage Theory and Social Complexity*. London: Bloomsbury Academic.

Deleuze, Gilles, and Felix Guattari. 1987. *Thousand Plateaus: Capitalism and Schizophrenia*. Translated by Brian Massumi. Minneapolis, MN: University of Minnesota Press.

Delormier, Treena, Katherine L. Frohlich, and Louise Potvin. 2009. Food and Eating as Social Practice—Understanding Eating Patterns as Social

Phenomena and Implications for Public Health. *Sociology of Health & Illness* 31 (2): 215–228. https://doi.org/10.1111/j.1467-9566.2008.01128.x.
Flegr, Jaroslav. 2007. Effects of Toxoplasma on Human Behavior. *Schizophrenia Bulletin* 33 (3): 757–760. https://doi.org/10.1093/schbul/sbl074.
Ginn, Franklin. 2014. Sticky Lives: Slugs, Detachment and More-than-Human Ethics in the Garden. *Transactions of the Institute of British Geographers* 39 (4): 532–544. https://doi.org/10.1111/tran.12043.
Halkier, Bente, and Iben Jensen. 2011. Methodological Challenges in Using Practice Theory in Consumption Research. Examples from a Study on Handling Nutritional Contestations of Food Consumption. *Journal of Consumer Culture* 11 (1): 101–123.
Haraway, Donna J. 2008. *When Species Meet*. Minneapolis, MN: University of Minnesota Press.
Hitchings, Russell, and Shu J. Lee. 2008. Air Conditioning and the Material Culture of Routine Human Encasement: The Case of Young People in Contemporary Singapore. *Journal of Material Culture* 13 (3): 251–265. https://doi.org/10.1177/1359183508095495.
Maller, Cecily. 2017. Epigenetics, Theories of Social Practice and Lifestyle Disease. In *The Nexus of Practices: Connections, Constellations, Practitioners*, ed. Allison Hui, Theodore R. Schatzki, and Elizabeth Shove, 68–80. Abingdon, Oxon: Routledge.
———. 2018. *Healthy Urban Environments: More-than-Human Theories, Routledge Studies in Environment and Health*. Abingdon, UK: Routledge.
Maller, Cecily, and Yolande Strengers. 2015. Resurrecting Sustainable Practices: Using Memories of the Past to Intervene in the Future. In *Social Practices, Intervention and Sustainability: Beyond Behaviour Change*, ed. Yolande Strengers and Cecily Maller, 147–162. Abingdon, Oxon: Routledge.
Medeiros, Denis, and Robert Wildman. 2011. *Advanced Human Nutrition*. Burlington, MA: Jones & Bartlett Learning.
Meloni, Maurizio. 2014. How Biology Became Social, and What It Means for Social Theory. *The Sociological Review* 62 (3): 593–614. https://doi.org/10.1111/1467-954X.12151.
Mol, Annemarie. 2002. *The Body Multiple: Ontology in Medical Practice*. Durham, NC: Duke University Press.
Oskar, Jenni G., and Bonnie B. O'Connor. 2005. Children's Sleep: An Interplay Between Culture and Biology. *Pediatrics* 115 (Suppl. 1): 204–216. https://doi.org/10.1542/peds.2004-0815B.

Pickering, Andrew. 1993. The Mangle of Practice: Agency and Emergence in the Sociology of Science. *American Journal of Sociology* 99 (3): 559–589.

Reckwitz, Andreas. 2002a. Toward a Theory of Social Practices: A Development in Culturalist Theorizing. *European Journal of Social Theory* 5 (2): 243–263. https://doi.org/10.1177/13684310222225432.

———. 2002b. The Status of the “Material” in Theories of Culture: From “Social Structure” to “Artefacts”. *Journal for the Theory of Social Behaviour* 32 (2): 195–217.

Rinkinen, Jenny, Mikko Jalas, and Elizabeth Shove. 2015. Object Relations in Accounts of Everyday Life. *Sociology* 49 (5): 870–885. https://doi.org/10.1177/0038038515577910.

Schatzki, Theodore R. 2002. *The Site of the Social: A Philosophical Account of the Constitution of Social Life and Change*. University Park, PA: The Pennsylvania State University Press.

———. 2010. Materiality and Social Life. *Nature and Culture* 5 (2): 123–149. https://doi.org/10.3167/nc.2010.050202.

Shove, Elizabeth. 2017. Matters of Practice. In *The Nexus of Practices: Connections, Constellations, Practitioners*, ed. Allison Hui, Theodore R. Schatzki, and Elizabeth Shove, 155–168. Abingdon, Oxon: Routledge.

Shove, Elizabeth, and Mika Pantzar. 2005. Fossilisation. *Ethnologia Europaea* 35: 59–63.

———. 2007. Recruitment and Reproduction: The Careers of and Carriers of Digital Photography and Floorball. *Human Affairs* 17: 154–167.

Shove, Elizabeth, Mika Pantzar, and Matt Watson. 2012. *The Dynamics of Social Practice: Everyday Life and How It Changes*. London: SAGE.

Strengers, Yolande, and Cecily Maller. 2012. Materialising Energy and Water Resources in Everyday Practices: Insights for Securing Supply Systems. *Global Environmental Change* 22 (3): 754–763. https://doi.org/10.1016/j.gloenvcha.2012.04.004.

Strengers, Yolande, Larissa Nicholls, and Cecily Maller. 2016. Curious Energy Consumers: Humans and Nonhumans in Assemblages of Household Practice. *Journal of Consumer Culture* 16 (3): 761–780. https://doi.org/10.1177/1469540514536194.

Taylor, Brian. 1993. Unconsciousness and Society: The Sociology of Sleep. *International Journal of Politics, Culture, and Society* 6 (3): 463–471.

Thrift, Nigel. 2008. *Non-Representational Theory: Space, Politics, Affect*. Oxon: Taylor & Francis.

Turnbaugh, Peter J., Ruth E. Ley, Micah Hamady, Claire M. Fraser-Liggett, Rob Knight, and Jeffrey I. Gordon. 2007. The Human Microbiome Project. *Nature* 449 (7164): 804–810.

Wainwright, Steven P., and Bryan S. Turner. 2006. 'Just Crumbling to Bits'? An Exploration of the Body, Ageing, Injury and Career in Classical Ballet Dancers. *Sociology* 40 (2): 237–255. https://doi.org/10.1177/0038038506062031.

Walker, Gordon. 2015. Beyond Individual Responsibility: Social Practice, Capabilities and the Right to Environmentally Sustainable Ways of Living. In *Social Practices, Intervention and Sustainability: Beyond Behaviour Change*, ed. Y. Strengers and C. Maller, 45–60. London: Routledge.

Wallenborn, Grégoire. 2013. Extended Bodies and the Geometry of Practices. In *Sustainable Practices: Social Theory and Climate Change*, ed. Elizabeth Shove and Nicola Spurling, 146–164. Oxon: Routledge.

Wallenborn, Grégoire, and Harold Wilhite. 2014. Rethinking Embodied Knowledge and Household Consumption. *Energy Research & Social Science* 1: 56–64. https://doi.org/10.1016/j.erss.2014.03.009.

Warde, Alan. 2014. After Taste: Culture, Consumption and Theories of Practice. *Journal of Consumer Culture* 14 (3): 279–303. https://doi.org/10.1177/1469540514547828.

Whatmore, Sarah. 2002. *Hybrid Geographies: Natures Cultures Spaces*. London and Thousand Oaks, CA: SAGE Publications.

Yong, Ed. 2016. *I Contain Multitudes: The Microbes Within Us and a Grander View of Life*. London: The Bodley Head.

6

Mobile Drinking: Bottled Water Practices and Ontological Politics

Gay Hawkins

Introduction

In her groundbreaking analysis of the ways in which different medical specialisms variously configure a disease and the body that suffers it, Annemarie Mol foregrounds the relationship between objects and the realities they enact. Central to this relationship are practices:

> If practices are foregrounded there is no longer a single passive object in the middle waiting to be seen from the point of view of seemingly endless series of perspectives. Instead, objects come into being—and disappear—with the practices in which they are manipulated. And since the object of manipulation tends to differ from one practice to another, reality multiplies. (Mol 2002, 5)

G. Hawkins (✉)
Institute for Culture and Society, Western Sydney University, Sydney, NSW, Australia
e-mail: G.Hawkins@westernsydney.edu.au

C. Maller, Y. Strengers (eds.), *Social Practices and Dynamic Non-Humans*,
https://doi.org/10.1007/978-3-319-92189-1_6

This chapter takes up this approach to investigate the rise and political effects of a new practice: carrying and sipping water from plastic bottles whilst on the move in urban space. From the mid-1990s, disposable plastic bottles of water began to appear everywhere, on desks in workplaces and universities, poking out of backpacks on public transport, in the hands and mouths of people as they walked down the street. No longer a device to support the fitness activities of a minority, the bottle of water was increasingly normalised as a personal accessory essential to enabling constant sipping on the move—or 'hydration support' as the marketing liked to call it. How did this new mobile practice impact on conventional infrastructural drinking water practices? Was the portable plastic bottle of water, and the constant sipping it enabled, just a trendy new fad and segment in beverage markets entirely unconnected to reticulated forms of water provision? Or was it implicitly interfering with these arrangements and the practice of drinking water from taps? How does a mundane disposable thing like a plastic bottle acquire the capacity to reconfigure how and where drinking happens and who should be responsible for providing drinking water?

These questions foreground the relationship between practices, material devices and politics. In pursuing them my aim is to get beyond existing activist critiques of bottled water and open up another approach. Many analyses of this new mobile drinking practice show little interest in its history or how it was practically assembled. Instead, the tendency is to ridicule the sipper or demonise the bottle. In the numerous activist campaigns that have emerged to oppose branded bottled water as an unsustainable commodity, it is often not the drinking practice that is under attack but the device for delivering the water: a single-use plastic bottle. It is seen as a passive object of corporate manipulation: a tool for packaging and privatising water that has turned citizens into gullible consumers and contributed to a global plastic waste crisis. But what if—heeding Annemarie Mol quoted above—we removed the passive object from the centre of the story and considered, instead, how it comes into being through practices and how these practices provoke new realities and politics? Rather than see the plastic bottle of water as a pre-existing material element that supports a problematic new drinking practice, what would it mean to consider how bottles come to matter and *why* in the emergence and activity of mobile drinking: how they are accounted for and enacted in practices—and with what effects?

Mol's argument is explicitly ontological and political. She insists that it is only through practices that we can understand objects and the way they are 'done'. However, she goes on to claim that in appreciating sociomaterial practices and the multiple realities they make, it is also necessary to acknowledge how these realities interact with other related ones and how they are intertwined and often exist in tension because, as she insists: 'if reality is multiple, it is also political' (Mol 2002, 7). This science and technology studies (STS) inflected approach to objects and practices has significant value for assessing the political effects of the rise of mobile drinking practices beyond activist critique. A focus on practice ontologies makes it possible to see how the emergence of mobile drinking is associated with existing ways of delivering and drinking water and how it diverges, converges or interferes with these. If the action of mobile drinking configures the plastic bottle not simply as a convenience commodity but as an alternative personal water supply, what does this mean for water realities delivered by networked pipes, and taps, and that involve public or private urban service organisations rather than beverage corporations and markets?

The relations between drinking practices enacted around disposable bottles and those associated with taps are at the heart of the analysis here. These relations cannot be reduced to a crude opposition between markets and states or private and public; many mobile drinkers do not understand or justify this practice in these terms. The plastic bottle of water is an unbounded object that simultaneously containerises water and makes it mobile. It also requalifies the meanings of water and drinking practices in complex ways with various effects. A focus on practices challenges the idea that there is a singular 'politics' of bottled water that can be read off the effects of this new conduct and that it is possible to critique these effects in advance. Instead, it is necessary to investigate how mobile drinking has emerged, what kinds of 'accountability relations' (Woolgar and Neyland 2013) sustain it and how this practice may implicitly or explicitly 'agence' (Hagberg 2016) the bottle and the water as objects of interference with other more sustainable drinking water practices.

In taking up this ontological approach, the focus here is primarily on the nature of mobile drinking practices and their contextualisation within beverage markets. This does not mean that mobile drinking is simply a product of corporate manipulation or that it can be considered exclusively as a 'consumption practice'. Consumer actions are one of several aspects under investigation. As a practice 'mobile drinking' involves a complex association of elements, materials, knowledges and performances—human and non-human, economic and non-economic. Beverage markets are a potent force in provoking this practice, but they don't control it. Drinkers use bottles in various unexpected ways far beyond what corporations envisaged or that marketing might suggest. Refilling the single-use bottle rather than buying another is one potent example of this. In the same way that plastic bottles and the water they contain reconfigure drinking, thirst, water qualities and the drinker into myriad new associations. Mobile drinking practices are configurations that both realise and exceed markets and consumption. And it is in these excesses, in the emergent realities that are enacted in the practice of constant sipping, that the political effects of practices can be detected and assessed.

In the rest of this chapter, I want to pursue two key questions: how did the practice of mobile sipping emerge and what specific attributes and accounts of the plastic bottle, the water and drinkers were configured in this process? And secondly, how are bottled water practices being *re*configured from a mobile activity to an alternative form of mass infrastructure? What might be the effects of this reconfiguration on utility-based sources of water supply and the idea of infrastructure as an unbounded network that generates a service relation with users by providing continuous access to water, rather than delivering it as a fixed volume commodity which the bottle does? As outlined, these questions foreground the ontological and political effects of practices—issues that are too rarely examined in much practice theory. Before pursuing them it is necessary to briefly review current debates about the relationships between practices, materials and politics.

Practice Theories, Materials and Politics

While the relations between practices and materials have been a prominent focus in much practice theory, there has been far less attention paid to how these relations may provoke or become caught up in political processes. Within the practice theory characterised by the work of Elizabeth Shove and Ted Schatzki, for example, there has been a valuable debate about the role of materials and resources in configuring practices. In Shove et al.'s (2012) book *The Dynamics of Social Practice*, materials and things are recognised as the stuff out of which practices are realised. Objects do not reflect society, they are not passive elements used to resource practices. Rather, they are central to constituting practices, and, at the same time, their particular material and technical capacities are constituted *in* practice.

The debt to Latour and ANT is acknowledged in this analysis. Shove and colleagues recognise the central importance of this field in foregrounding the materiality of the social. Materials and objects have strategic status within practice theory where they are mediators and co-producers of practices. However, practice theory also diverges from ANT and STS because it rejects the claim that materials and objects are—supposedly—immutable carriers of power and influence (Shove et al. 2012, 10). Shove et al. also critique ANT for having a fairly leaden view of objects that is unable to effectively understand exactly what they do in practices and how they are configured in association with numerous other elements, human and non-human:

> actor network theory has inspired politically and philosophically intriguing debates about the relation between humans and non-humans with whom they share their lives, but has ironically done so in ways that divert attention from more ordinary questions about what these cyborg/hybrid entities are actually *doing*. (Shove et al. 2012, 10)

This critique unfairly misrepresents ANT and STS analyses of the non-human and the dynamics of assembling the social. These issues have been an ongoing concern in these fields, and recent work has been particularly suggestive for reconfiguring the concept of 'practices' and the relationship between materials and political processes. Mike Michael classifies this recent work as the 'ontological approach in post-ANT' (2017, 120). The work of Mol is central here and also Woolgar and Neyland's (2013) important book *Mundane Governance*. These authors are interested in practices and their role in the ontological enactment of realities. However, in the shift from a focus on practices as the central object of analysis to a focus on the *relationship* between practices and ontology—or how various realities are practically enacted and accounted for—a range of concerns emerge that are somewhat different from the usual preoccupations of practice theory. These are the dynamics by which various ontologies or versions of reality interact, the ways in which ontologies are accounted for and become seen as normal or unexceptional and the processes whereby things might come to do political work. A key point is that practices are not the organising category in this STS approach. 'Practices' are used descriptively and investigated empirically, to explain a wide variety of co-ordinated actions that help make or enact realities.

Woolgar and Neyland's (2013) detailed studies of the micro-politics of everyday life offer powerful insights into how objects acquire various capacities: technical, moral, political and more. Their central focus is with how these capacities come to shape conducts and realise forms of governance or the conduct of conduct. In briefly reviewing this work, the aim is to show how STS-inflected analyses can get far closer to understanding the complex interactions between materials, practices and politics than much mainstream practice theory.

According to Woolgar and Neyland (2013), one of the key achievements of STS accounts of objects and technologies is the powerful (and oft repeated) claim by Latour that technology is society made durable. This claim—often crudely represented as technology expresses or reflects social forces—is the basis for arguments about social construction, or more specifically the assertion that social forces shape or guide the development of the technological. While these arguments are valid, they do not capture the complexity of STS debates about the relations between the social and the technical, and the ways in which these relations are shaped by practices and might become political.

Woolgar and Neyland claim that while many STS authors refuse to see technological effects as essential and neutral, thereby demolishing the assumption that they have fixed capacities, they often fall into the trap of shifting their essentialist reasoning to the social and political: 'the presumption is that, whereas we cannot straightforwardly read-off the technical characteristics of the technology, we are somehow capable of reading off the (actual) social and/or political characteristics that apply' (2013, 39). To counter this tendency to replace technological determinism with social determinism or social construction, they develop an argument and method for understanding the connections between objects and *ontology*, or the forms of accountability and reasoning that explain what things are and what they should do in the world and what they are responsible for. In their analysis, determinations of the ontological status of things are always contingent and always dependent on practices.

The issue is how are various object ontologies formatted and contested? Woolgar and Neyland's (2013) answer to this question is to focus on how specific 'accountability relations' emerge and take hold. At the most basic level, accounts involve explanations about what something is and what it is capable of. However, accounts also often involve subtle moral justifications about what things are answerable for and how they should be responded to. In this way, the technical and the social become inextricably mixed up. Accountability relations are also constituted in action, through social practices what is considered accountable becomes a 'mundane, pervasive, organizing orientation' (Woolgar and Neyland 2013, 32). There are also, of course, formal and official mechanisms for giving accounts or justifying actions but even these are dependent on 'routine, on moment to moment interaction through which sense is made of the system and accountability established' (Woolgar and Neyland 2013, 32). Central to this important argument is the insistence that the practices and processes whereby objects and things are made real and accountable are fundamentally political. For Woolgar and Neyland (2013), politics is constituted at the level of ontology, in the processes of deciding what will count and be accountable in this context and how? Politics is not simply the preserve of traditional large-scale organisations, policies and structures, it is thoroughly entangled with the mundane and the everyday and the ways in which practices involve very particular accounts of things and enact realities.

Another important and related analysis of the relations between objects, ontologies and politics comes from French sociologist Laurent Thévenot. In his essay 'Which Road to Follow? The Moral Complexity of an Equipped Humanity', Thévenot (2002, 54) explores the nature of what he calls 'pragmatic engagements with the world'. Thévenot is interested in understanding how objects and technical devices participate in social practices and, like Woolgar and Neyland (2013), how they become implicated in moral and political orders. This is not a question of humans projecting onto objects moral justifications or political concerns. This assumes that humans are autonomous individuals with an already equipped moral agency that is directed to objects. Rather, through everyday actions humans constantly single out and evaluate what is relevant and pragmatic, and the functional agency of objects, their technical capacities to do things in the world, is implicated in shaping these evaluations. As Thévenot says: 'The moral agency of humans is achieved only with the support of other elements' (2002, 59). It is not that humans don't bring to these pragmatic encounters already existing 'regimes of engagement' and orders of worth, rather that these are constantly being tested and qualified in use, in everyday practices. Different regimes of engagement provide different kinds of access to making sense of reality or different 'reality tests'. Of most interest here is Thévenot's (2002) notion of 'familiar engagement' where people invest in their environment through continued use, customs and routines in order to generate feelings of convenience, comfort and ease with their surroundings.

Thévenot (2002) shows how people and objects are caught up in complex processes of evaluation in which what counts as 'good' and 'real' become linked through the dynamics of everyday pragmatic engagements with the world. New objects then, like the emergence of 'bottled water', had to withstand reality tests. This product and object had to be evaluated and qualified in terms of what was this good for? Marketing was a critical official site where such accounts were provided (hydration on the move, health, avoiding risky tap water) but so too were everyday practices where drinkers accommodated the bottle into their own vernacular daily activities. Another imperative for reality tests is to generate shared or general justifications for objects so that they can have the effect of co-ordinating common actions and practices, so that objects and uses can be normalised. Carrying bottled water and engaging in mobile drinking was often a public practice, done on the street or in the office in full visibility, this aspect of the drinking practice no doubt facilitated the sense that it was acceptable: everyone was doing it!

A primary focus in Woolgar and Neyland's (2013) and Thévenot's (2002) arguments is how regimes of justification and accountability relations for things are mobilised and become durable. The essential contestability of reality means that deterministic or structuralist analyses are impossible to sustain. Instead, the issue is how are various realities formatted and contested? How do practices make sense of things and how do things make sense of practices? In moving now to an investigation of the rise of mobile drinking, this 'ontologically sensitive' analytical framework from recent work in STS is central. Not only does it foreground the importance of ontological constitution and enactment in making this practice both mundane and political. It also shows that it was impossible for this practice to spread and become normalised without the development of diverse accountability relations or justifications that established the meaning and purpose of bottles and mobile drinking.

Inventing Mobile Drinking: Discourses and Devices

How did the carrying of a disposable bottle of water come to be seen as a necessary resource for urban life and mobility? In what ways can we understand constant sipping on the move as a new drinking practice? One way to answer these questions would be to adopt a deficit model and point to the fact of diminishing access to free drinking water in cities. This approach assumes that the emergence of bottled water was largely a reaction to the disappearance of previous public forms of supply. Explaining changing water infrastructures and the steady demise of public water fountains over the last 30 years in many urban centres is beyond the scope of this chapter. However, it is generally linked to the inability of urban authorities to maintain these facilities, and the growth of vending machines and convenience stores selling numerous sweetened beverages, or 'fast-moving consumer goods' as they are classified by the industry. That said, the idea that there is a causal relation between disappearing access to water in public space and the concomitant rise of a market in single-serve branded bottles is difficult to sustain. This mode of analysis reduces complex changes in everyday practices to the opportunistic logic of the corporation and decline in water infrastructures and state provision. It equates market formation with the privatisation or commodification of water and fails to understand how market processes are embedded in wider cultural dynamics and how they are practically done.

Empirical research for our book *Plastic Water* (Hawkins et al. 2015) showed that the carrying of bottled water was much more connected to shifts in consumer concerns about health and water quality than the disappearance of public water fountains. Carrying a personal supply of water was generally not a response to reduced public provision in urban space; it was a response to the reconfiguration of the relationship between drinking and the mobile body. Carrying water so that it is always available is very different from opportunistically grabbing a drink at a water fountain. This frequent sipping practice assumed constant self-surveillance of the body and the regulation of fluid intake. This was not the same as deciding to quench your thirst if you happened to encounter a water fountain.

This new carrying practice was also connected to the bottle itself, to the ways in which it emerged an indispensable resource in the enactment of a new mode of drinking and new water realities. This claim may sound banal—obviously you cannot carry water without a container—but the materiality of the bottle mattered. The nature of the container played a crucial role in formatting the carrying practice. STS refers to this as 'scripting', but I want to extend this account by arguing that the container did more than simply direct or shape practices—it also functioned as a 'market device' central to requalifying the values and meanings of water and the drinking subject. As Muniesa et al. (2007) argue, market devices *do* things: they help articulate economic actions in relations with other devices. Market devices also help constitute distinct consumer practices and nowhere is this more obvious than in the case of the plastic container. For Cochoy and Grandclément-Chaffy (2005), consumer packaging is a key interface between production and consumption: it has to facilitate the efficient framing and distribution of the product *and* it has to appeal to and configure the existing habits of the consumer. In this dynamic interaction, packaging transforms both the product and the consumer. As I have argued elsewhere, the concept of a market *device* foregrounds the performativity of packaging, its capacity to interact with products and consumers and provoke changed social relations between them that become durable (Hawkins 2012).

How then did carrying bottled water become implicated in enacting new codes of self-care, in shaping a new drinking reality and experience of thirst? And what role did developments in packaging play in this process, particularly the invention of the PET bottle? Obviously, there are many other elements involved in the practical and ongoing enactment of mobile drinking, but these two elements foreground the critical role of accountability relations and material devices in establishing a new practice and drinking reality.

Reconfiguring Thirst and Drinking Bodies

In a key article on the rise of markets in bottled water, Kane Race (2012) has argued that the formation of this market depended on assembling a particular subject: 'the subject of hydration'. Central to his analysis is the assertion that the bodily experience of thirst has been reconfigured by the practice of constantly sipping water on the move. The drinking body in bottled water practices is a body that has 'learnt to be affected' (Latour 2004, 205) by thirst in new ways. Carrying a bottle of water does not simply address the fixed biological needs of the body it actively generates and reorders them by provoking the body to become used to drinking more water more often.

Using historical research Race (2012) traces how the emerging fields of health, fitness and exercise science in the 1970s played a key role in producing new scientific facts and principles about hydration. As sport became more professionalised, researchers started to investigate the effects of physical exercise on bodies and athletic performance. One of the main issues researched was dehydration and the implications for marathon runners of drinking, or not drinking, during a race. Via a range of experiments, it was found that taking constant regulated amounts of fluid during extreme exertion aided performance. These findings prompted a major revision in marathon rules as well as changes in training. They also put 'hydration support' on the agenda of exercise and sports science as a new focus for research.

However, the key question Race (2012) pursues is how did this very situated biomedical research move out of the laboratory and the running track into general discourse; how did hydration become a popular everyday concern? The answers to this question are varied, but Race (2012) links the main momentum to the growth of jogging and fitness cultures throughout the 1980s. The spread of these practices often involved the promotion of the new science of hydration. In caring for the body during exercise, adequate fluid intake was promoted as an essential form of self-regulation. As Race argues: 'in this moment, the question of fluid intake is disconnected from the body's experience and rendered a matter of expert calculation and determination … Thirst is not what it used to be' (2012, 84–85). Race's research shows how the reconfiguration of thirst into 'hydration science' prompted new forms of self-surveillance based on concerns about water and what it was now considered to be responsible for. Biomedical and scientific discourses reframed drinking water from an ordinary biological need into a medical concern. This mundane liquid material and the activity of consuming it frequently was now directly implicated in health.

Equally significant was how people learnt to become affected by these discourses and incorporate them into everyday ways of conceiving and acting on the self (Race 2012, 73). Initially, the practice of carrying water and frequent sipping was largely restricted to those doing exercise. The spread into the wider population, marked by the carrying of bottles of water during ordinary activity, happened later. This move was largely aided by the marketing practices of beverage markets. Picking up on the rise of biomedical languages in food and pharmaceutical marketing, and evidence of a growing population of health-conscious consumers, beverage companies began using hydration science to sell water as a new product line. Bottled water marketing often suggested that everyone needed to ensure they were getting 8 × 8 ounce glasses of water per day. The only way this goal could be realised was by carrying water with you *at all times* so that you could sip regularly. In this way, the bottle was promoted as an essential item for those interested in a 'healthy lifestyle'—mostly women. Race's argument shows how the science of hydration was translated into consumer information and used to encourage a new drinking practice in the wider population. Of course, not everyone took up this practice, but there is no question that the rapid growth of markets in bottled water during the 1990s could have happened without the medicalisation of thirst and the emergence of popular concerns about hydration requirements. This complex process of developing new accounts of water and drinking, of reconfiguring thirst into 'hydration requirements', enabled frequent sipping to be disconnected from extreme exercise and popularised as an ordinary act of self-care and personal vigilance. Healthy bodies needed to carry water so they could maintain hydration needs anywhere anytime.

Devices: Bottles and Brands

If constantly accessing water had become a personal health technique, then enacting this technique, and responsibility to the self, required devices. This is where the PET bottle becomes central. Drinking on the move required containers, and the rapid adoption of PET bottles in the beverage industry from the early 1980s onwards proved to be crucial. As argued elsewhere (Hawkins et al. 2015), PET transformed the beverage industry by rapidly displacing glass and aluminium cans as the preferred packaging material. This plastic was tougher than glass but had equal translucence and clarity; it was also incredibly light making movement and distribution over long commodity chains cheaper. More important, however, was the impact on consumers. Here was a container that invited portability because of its remarkable durability and lightness. These key material affordances helped make PET a potent market device in the beverage industry. They also played a crucial role in the conversion of water into a market thing. For not only could the PET bottle of water be dropped, resealed and refilled, unlike the glass bottle or can, its clear flowing plasticity semiotically echoed and enhanced the liquid contents: water seemed purer in PET. The PET bottle was a central actant in changing the qualities of water and how consumers accessed and drank it. It was the perfect container technology for reformatting drinking water as a new health and carrying practice.

Lightweight plastic packaging, from polystyrene to PET, has been central to the expansion of the fast food industry and eating in public places. Most of this packaging is for single-use items; its function is to facilitate consumption on the move. In the case of the PET bottle of water, however, the packaging enabled a new practice of consumption *and* carrying. The bottle of water was often not purchased for immediate consumption on the spot but for constant sipping throughout the day. In this way, the bottle of water defied its status as disposable or fast moving packaging and became an important personal accessory: something to put in the backpack or handbag along with the wallet and mobile phone and something to have on hand at all times. As this practice increased, some backpacks and handbags began to incorporate bottle holders into their design, a sure indicator that the bottle had become normalised as an essential everyday accessory. Significantly, as anti-bottled water campaigns escalated in the twenty-first century in response to the waste impacts of this practice, new markets emerged selling refillable bottles in various materials. These markets were an opportunistic response to activism; they recognised and rematerialised an existing practice. Even so, markets for single-use PET-bottled water continue to grow, and today it is the largest beverage market in the USA, overtaking sweetened beverages.

Another central element connected to the package was branding. STS-inflected economic sociology has shown that branding is a critical device in the assemblage and co-ordination of new markets. Brands are platforms for the organisation of value, and one of their key functions is not only to differentiate the product from other similar ones but also to establish its unique qualities. According to Callon et al. (2002, 201), central to all markets is an 'economy of qualities' that is produced through the processes of qualifying and positioning goods: 'Through construction, a product is always singular and similar to other products because it is immersed in a space of qualities that makes comparisons possible'. These dynamics of positioning and comparison occur not just between different bottled water markets and products but also between water as a commodity and water as an infrastructural service, between water as packaged and water that flows. In the case of bottled water, the work of the brand was to reveal to the drinker new meanings and values for water, to intensify its unique qualities. One of the common ways in which this was done was through elaborate celebrations of the water source: usually pristine, 'untouched', natural and more. This new account of water—where it had come from and how pure it was—implicitly devalued infrastructural water that was not framed in the same way, that appeared to be without elaborate qualities and from an unknown source. The key point about branding was that it was another important element in reconfiguring bottled water as healthy and risk-free, brands offered a new account of water sources and a new regime of value for this mundane liquid. Not only was water in bottles different, it was also better and safer. In this way branding amplified the health discourses of hydration 'science' and offered another justification for why drinking water from bottles was a more responsible drinking practice.

According to Mol (2002, 6), an 'ontology is not given in the order of things … instead, ontologies are brought into being, sustained or allowed to wither away in common, day to day material practices'. In the ontological enactment of mobile drinking, we can see how this emergent reality was partially provoked by new accounts of hydration. With the help of marketing by beverage corporations, these accounts migrated out of elite sports research into the everyday life of increasingly health-conscious populations where practices of self-regulation were already established. New discourses about drinking and hydration appealed to drinkers who had already learnt to be affected by health marketing. These new accountability relations justified the practice of constant sipping as a healthy activity, but they needed more than discourses and marketing to be realised. The role of the bottle was more than just a container technology, it was also central to enacting water as a market thing, something to be purchased and something that was branded. Carrying the bottle and sipping on the move was an urban practice that tested and extended the sociotechnical capacities of the bottle beyond disposable packaging. It also helped diffuse this new drinking reality into numerous public and institutional spaces. Sipping in public or at work classified the drinker as 'healthy' and also implicitly promoted the idea that there were alternative and possibly better waters than tap.

These are some of the key factors that came to count in mobile drinking and that helped make this practice ordinary and 'good'. The bottle of water did not have these essential qualities, they had to be done, and in the process of enacting them, the bottle and the water acquired new meanings in relation to each other and in relation to drinkers. They acquired the capacity to shape drinking practices in specific ways, to become a device that was now responsible for caring for and governing the body by reconfiguring the biological experience of thirst into hydration needs. These accountability relations for bottled water established a drinking practice that significantly diverged from those focussed on taps and occasional opportunistic sipping in public. But what of bottled water's political capacities: how did this new market-based drinking practice interfere with other forms of water and infrastructural modes of distribution?

Bottles and Ontological Interference

We have seen how mobile drinking established a new ontological reality for drinking that diverged from previous practices. In this section I want to examine two examples where this practice can be considered as a form of ontological interference rather than mere divergence or diversity in drinking practices. The idea of interference opens up questions about the politics of practices, how they can disrupt or undermine other more equitable or sustainable practices by enacting a different reality that subtly contests existing ones. In the first example, we see a familiar expression of politics as organised contestation and controversy via online activism. In the second we see a far more understated manifestation of ontological interference in which a bottled water marketing campaign promotes an alternative set of accountability relations for the bottle and what it can be responsible for. These promotional accounts for the bottle enact a new drinking and infrastructural reality that is very disturbing in its implications.

In April 2014 a controversy about the safety of the public water supply erupted in Flint, Michigan, in the USA. Due to long-term maintenance failure on the part of the local administration, the water pipes servicing hundreds of thousands of homes in poor communities were revealed to be delivering dangerously high levels of lead. The lead poisoning and neurological damage to all residents but particularly children was significant and irreversible. In response to huge national and international media coverage, various celebrities sent truckloads of bottled water as a gesture of support. In response to that response, Michael Moore, well-known celebrity activist, launched a major online campaign: 'Don't Send Bottled Water'. Moore's argument was that by turning to beverage markets to solve a problem of state failure, concerned celebrities were simply letting government off the hook. They were allowing the state to abdicate its responsibility to supply safe drinking water to citizens. This well-meaning gesture from celebrities effectively endorsed an unsustainable product and positioned branded bottled water and corporations as an alternative to infrastructure. Rather than turning to markets to solve this problem of significant state failure, Moore advocated the arrest of the governor of Michigan.

The second example comes from a 2016 marketing campaign for Mt Franklin bottled water, Australia's leading bottled water brand and part of Coca Cola's suite of packaged waters. This public billboard and online campaign used the image of a huge bottle of Mt Franklin next to a map of Australia with the slogan 'Mt Franklin—the nation's hydration'. This technique of linking a product to the nation is nothing new. Everything from beer to beds has been promoted in the same way. But there was something slightly different here. Mt Franklin wasn't being qualified as a water that expressed national identity, as embodying what was uniquely Australian. It was being qualified as a water acting *in the service of the* nation. To position bottled water as 'the nation's hydration' was to implicitly frame it *as* a form of infrastructure, as a key element in the sociotechnical systems that both support the life of populations and constitute a national constituency. Note that the advertisement refers to the 'nation', not Australia. This bottled water campaign reconfigured mass water provision to populations in extractive rather than service terms. That is, it celebrated a mode of delivery based on extracting water and packaging and purchasing it rather than in terms of a service relation between the state and citizens in which water infrastructure and catchments are understood as a shared common resource and government responsibility.

According to Mt Franklin's marketing, keeping national citizens 'hydrated', once a core function of pipes and the state, was now possible using 250 millilitre branded PET bottles.

Both these examples can be considered as evidence of what Wendy Brown (2015) has termed 'neoliberal reason'—a process that converts the political character of democracy's constituent elements into economic ones. At an abstract or general level, this is a useful observation. However, it doesn't adequately capture the specific dynamics involved: the types of interactions, practices, materials and calculations that unfold between bottled water markets and other modes of water distribution and governance. In describing these processes as neoliberalism or privatisation, there is an implicit tendency to oppose state and markets rather than investigate *how* these categories are actively constituted, or done in practice, and how they interact and hybridise and how interference can characterise political processes.

In the case of Flint, Michael Moore's activist campaign openly contested a market on the grounds that it was undermining a state function. As the controversy played out, it manifested all the elements of a hot situation and the unfolding of 'issue politics' (Marres 2007). In the case of Mt Franklin, the political process and qualification of the market was less obvious. This was a piece of ubiquitous mass marketing so its affective registers were meant to be positive and motivational. Despite this, the campaign had an unsettling resonance, a sense that an accepted order or reality was being challenged. By equating a market good with a national infrastructural service, other regimes of value and accounts of water governance were effectively displaced. This undermining or interference was achieved by a subtle ontological sleight of hand. This mundane bottle of water invited a very different orientation and attachment for the consumer than other marketing campaigns for this ubiquitous product. Bottled water wasn't being promoted as an individual source of health and hydration, or as a fast-moving consumer good offering the convenience of water on the go, it was framed as a national service and utility. In this framing bottled water became a hybrid object *both* commodity and infrastructure that suggested new practices for accessing water in the home: substituting turning on the tap with reaching for a bottle. This bottle subtly reconfigured the existing accountability relations for mass water provision, and in this move it enacted what Woolgar and Neyland

(2013) refer to as 'ontogovernance', that is, new modes of justification for a mundane object that represent it as having the technical and moral capacity *to become responsible* for national water supply. This is ontological interference.

Conclusion

This ontologically sensitive account of mobile drinking and the rise of bottled water has shown how practices are implicated in making ontologies and configuring material things and thirsty bodies. Ontology is not an abstract philosophical concept that precedes or informs the social, it is located and enacted in practice. Central to the analysis presented here is the role of accountability relations in making things count and justifying what they are responsible for. These accountability relations shift as bottles move into different settings, associations and practices. These shifts provoke new accounts and realities, new bottles and new water. They can also prompt ontological interference through the subtle invocation of moral or socio-technical justifications that implicitly foreground the superiority of some practices over others, as the examples of Flint and Mt Franklin showed. These disturbing examples are evidence of the political effects of mobile drinking, the ways in which bottled water becomes capable of configuring a new water reality that appears 'better' than existing forms of infrastructural provision. As bottled water markets grow around the world and it is used increasingly in homes, not just on the move, it is possible to see an emergent water future. In this future—that for many is already here—the practice of reaching for the bottle, rather than turning on the tap, effectively reconfigures and displaces water provision from an unbounded service relation to an encounter with a bounded branded container.

References

Brown, W. 2015. *Undoing the Demos*. Cambridge, MA: The MIT Press.
Callon, M., C. Méadel, and V. Rabeharisoa. 2002. The Economy of Qualities. *Economy and Society* 31 (2): 194–218.

Cochoy, F., and C. Grandclément-Chaffy. 2005. Publicizing Goldilocks' Choice at the Supermarket: The Political Work of Shopping Packs, Carts and Talk. In *Making Things Public—Atmospheres of Democracy*, ed. B. Latour and P. Weibel, 646–657. Cambridge, MA: ZKM and MIT Press.

Hagberg, J. 2016. Agencing Practices: A Historical Exploration of Shopping Bags. *Consumption, Markets and Culture* 19: 111–132.

Hawkins, G. 2012. The Performativity of Food Packaging: Market Devices, Waste Crisis and Recycling. *Sociological Review* 60 (2): 66–83.

Hawkins, G., E. Potter, and K. Race. 2015. *Plastic Water: The Social and Material Life of Bottled Water*. Cambridge, MA: The MIT Press.

Latour, B. 2004. How to Talk About the Body? The Normative Dimension of Science Studies. *Body and Society* 10 (2–3): 205–229.

Marres, N. 2007. The Issues Deserve More Credit: Pragmatist Contributions to the Study of Public Involvement in Controversy. *Social Studies of Science* 37 (5): 759–781.

Michael, M. 2017. *Actor-Network Theory—Trials, Trails and Translations*. London: Sage.

Mol, A. 2002. *The Body Multiple: Ontology in Medical Practice*. Durham, NC: Duke University Press.

Muniesa, F., Y. Millo, and M. Callon, eds. 2007. *Market Devices*. Oxford: Blackwell.

Race, K. 2012. Frequent Sipping: Bottled Water, the Will to Health and the Subject of Hydration. *Body and Society* 18 (3–4): 72–98.

Shove, E., M. Pantzar, and M. Watson. 2012. *The Dynamics of Social Practice*. London: Sage.

Thévenot, L. 2002. Which Road to Follow? The Moral Complexity of an Equipped Humanity. In *Complexities—The Social Studies of Knowledge Practices*, ed. J. Law and A. Mol. Durham, NC: Duke University Press.

Woolgar, S., and D. Neyland. 2013. *Mundane Governance—Ontology and Accountability*. Oxford: Oxford University Press.

7

Immersed in Thermal Flows: Heat as Productive of and Produced by Social Practices

Elspeth Oppermann and Gordon Walker

Introduction

'Heat' is used to refer to a multiplicity of sensations, properties, energetic sources, and manifestations. It conjures both enjoyable and dangerous visual and somatic experiences of warmth and energy. When understood through thermodynamics and thermal physiology, heat is the key register of being and emergence: embroiled in all that is human and non-human, it plays a foundational role in the basic transactions, flows, and formations of the universe, from the 'big bang,' to exchanges within ecosystems, to the tiniest metabolisms of energy in the body (Ong 2012; Jablonski 2013). While heat is always present and always in-motion in these accounts, heat seems to matter to socially oriented accounts of the

E. Oppermann (✉)
Northern Institute, Charles Darwin University, Darwin, NT, Australia
e-mail: Elspeth.Oppermann@cdu.edu.au

G. Walker
DEMAND Centre (Dynamics of Energy, Mobility and Demand),
Lancaster University, Lancaster, UK
e-mail: g.p.walker@lancaster.ac.uk

C. Maller, Y. Strengers (eds.), *Social Practices and Dynamic Non-Humans*,
https://doi.org/10.1007/978-3-319-92189-1_7

world only when particular excesses or deficits of heat render these background flows conspicuous or materially compelling. Such contrasts challenge us to ask how heat should be made sense of, conceptualised, and accounted for in schemes of analysis that seek to do justice to more-than-human ontologies.

Our concern in this chapter is specifically to engage with how heat can be conceptualised in theories of social practice. In doing so we add to the array of 'non-humans' that feature across the chapters of this book, extending beyond other species or objects, to include energetic-material relations, specifically discussed here as thermal flows. Furthermore, we argue that heat has dynamic qualities that do not support simply adding it to a list of material 'elements' that are integrated and held together in and through the performance of social practice (Reckwitz 2002; Shove et al. 2012). To do so is to render it a relatively discrete and stable object or entity. Instead, heat is better understood as a dynamic energy within which all social practices are immersed, transgressing the human/non-human distinction and inherently both productive of practices and produced *by* practices (for related lines of argument regarding the biological rather than the energetic, see Maller 2018 and also Rinkinen 2018). This dynamism means that energetic-material relations, and potentially other related phenomena, can be added to the recent conceptual opening up of various forms of interrelation and interconnection between practices (Hui et al. 2017), as anticipated and experienced flows weave their way between and through complexes of practice performances.

We developed this understanding in response to outdoor workers' somatic experience of heat in northern Australia. Before focussing on this setting, we first draw on scientific literatures on thermodynamics and thermoregulation, then connect them to Schatzki's (2010) conceptual distinction between social practices and material arrangements as a way of entering into understanding outdoor manual workers' experience and management of extreme heat conditions. Schatzki's alertness to the diversity of material entities within which practices transpire—including forms of biological and physical flow that "pass through practice-arrangement nexuses" (ibid., 137)—provides a route into conceptualising heat in social practice and drawing out the particularities and complexities of its imbrications with practice performances. To explore these ideas further, we then examine how heat is produced by, and also productive of, the work practices of outdoor manual workers. This enables us to reveal the micro-temporalities and spatialities through which heat flows. Such flows are productive of practices, but practices also produce, move, and change thermal flows. As such the thermoregulation of human bodies occurs within a wider 'economy' of heat in the environment and broader thermodynamics, which practices both respond to and modulate.

The Materiality of Heat: Thermodynamics, Thermoregulation, and Immersive Flows

Two key scientific ways of knowing heat are thermodynamics and thermal physiology, specifically thermoregulation. These are realist modes of speaking about human-non-human relations of bodies, matter, and energy and how heat relates to these forms through constant, dynamic, physical circulations and flows. Thermal physiology and thermoregulation feature in our discussion because bodies are necessarily present in all social practices (at least as conventionally understood—see Maller 2018, Chapter 5) and because they are central to the empirical case we explore. However, human bodies are only one of a host of entities through which heat flows and, as we will make clear, are not in any sense independent of these wider-ranging and dynamic relations.

In thermodynamics, heat is understood as circulating through energetic exchange. The entire universe, and our particular world within this, is seen as composed of stocks and flows of heat and the changes of state they enable. It is the maintenance of the economy of these thermal flows on a planetary scale that is held to create the homeostatic envelope upon which all life depends (Ong 2012). Human bodies, and, for Bataille (1988), the entire human condition, are immersed in thermodynamic flows of energy extending from and out into the universe, flows that are largely taken for granted and invisible apart from in their consequences, particularly at moments of modulation and change. Indeed, evolutionary biologists argue that human bodies have evolved in direct relation to the thermal radiation arriving on the planet from the sun (Jablonski 2013) such that the emergence of skin as bodily surface is in part to manage the heat absorbed and retained by the body. Human bodies could be said then to literally embody the historical co-production of matter and energy and are constantly engaged in managing thermodynamic flows as part of ensuring their ongoing, successful functioning.

Thermal physiology identifies 'involuntary' and 'voluntary' ways in which humans 'orchestrate' their own thermoregulation, responding both to the heat in their external environment, such as from solar radiation, and the heat bodies produced through metabolism. How our bodies regulate blood flow, metabolism, and sweating are all involuntary processes that react to each other and our environment to keep our core temperature at around 37 °C, essential to keeping organs functioning (Jablonski 2013). When involuntary mechanisms are inadequate to maintain an ideal core temperature, voluntary or 'conscious' techniques of managing heat are said to come into play. These include actions or behaviours such as drinking something warm or cold, donning or removing items of clothing, or engaging in physical exertion to release energy as heat from muscular motion in order to warm up the body (Jablonski 2013, 52). The need for such responses indicates that thermoregulation does not take place in a vacuum; how hot the body gets is an effect of the dynamic relation between human bodies and their environments, broadly conceived. Solar radiation, air temperature, humidity, wind speed, and contact with differently convective substances like air, water, metal, and fabric all affect how efficient the body is at losing, gaining, or maintaining heat.

So what do such scientifically rooted understandings of thermal flows within and beyond the body mean for how we might think about heat in social terms? Our profound inhabiting of thermodynamics leads Ong (2012) to question ontologically where our bodies end and the dynamics of heat begin; "Heat is like space. [...] Space enables movement and experience but it is defined by the objects … which frame it" (Ong 2012, 15), such that heat is not a thing or object but is discernible in them and the relationships between them. Ingold (2007) sets up a similar framing in his exploration of weather, knowable not as events or by objects but through our sensory relationship with it; "persons and things relate not as closed forms but by virtue of their common immersion in the generative fluxes of the medium—in wind and weather" (Ingold 2007, 1). Like Ong, Ingold (2007, 1) transgresses and indeed repudiates the human-non-human boundary here, illustrated with the example of how respiration "continually disrupts any boundary between earth and sky." Following Ingold and Ong, thermoregulation similarly dissolves the boundary between human and environment, allowing for a dynamism of energy and matter which pervades, precedes, and exceeds the human. This is not to say the human does not matter, but to say it matters in the sense that it is part of these flows, and can make itself matter (sometimes literally) by modulating the circuits of thermal exchange where it has the ability to do so. We note, of course, that there are some generative complexes of thermodynamics which bypass or vastly exceed such abilities.

As such, heat is 'elemental' in the medieval sense of a dynamism that pervades the 'mediums' of the world, an energetic contingency, glimpsed through different knowledge traditions, from the scientific to the somatic. As we shall argue in the next section, an elemental reading of heat is helpful in extending our understanding of human-non-human relations in social practices, as it allows for material objects (including bodies) and activities to be placed in a co-constitutive relationship of matter-energy (Clark Forthcoming). The human itself, as an assemblage of materialities, is saturated in, dependent on, and produced by flows of energies, including of heat.

Elemental Heat and Social Practices

Armed with this elemental account of heat, we can now consider how heat can be made sense of in the burgeoning body of work on social practices. A common starting point is Reckwitz's (2002, 249) account of social practices as organised configurations of heterogeneous elements that are interconnected to one another:

> *...a routinized type of behaviour which consists of several elements, interconnected to one other: forms of bodily activities, forms of mental activities, 'things' and their use, a background knowledge in the form of understanding, know-how, states of emotion and motivational knowledge.*

Others have reworked these elements into alternative categories, such as meanings, skills, and materials (Shove et al. 2012; Shove and Spurling 2013). However, if we follow this element-oriented conceptualisation of social practices, we immediately encounter problems with including heat. What type of element could heat be? Is it a 'thing' in Reckwitz's scheme, somehow equivalent to artefacts and other material elements that are integral to, say, the practices of cycling (bicycle, bike lamp, pathway)? What of the energy that enlivens and moves *through* but also *on* from these things? Furthermore, heat, like other forms of energy, is also *required* for and *produced* by some practices, whether as kinetic energy (to push bike pedals) arising from metabolism and producing heat in the process. If heat is an energetic dynamism required for and produced through practices, it seems difficult to sensibly include it in a framework based on identifying constitutive practice elements.

It is better to see heat not as *an* element in the sense of being a discrete object, rather, as we have already argued, it is element*al* in the sense of being a basic or fundamental force. Heat in these terms is an energy that circulates within and between 'the' elements or components of practices, enabling their very materiality to mobilise and emerge at a chemical and physical level. As such, heat is never still and never fully contained. Indeed it is this inherent dynamism that makes it so valuable in enabling and mobilising practices as well as life and matter more broadly.

Some degree of recognition of these characteristics of flow is provided by Schatzki (2010). For Schatzki, not everything can be included in a listing of elements. He analyses social phenomena as "slices or aspects of nexuses of practices and material arrangements" (Schatzki 2010, 123), thus separating material arrangements from practices at the ontological level. Practices, he argues, are chains of actions, which are produced by and productive of material configurations, perpetually entwined or meshed with material arrangements. He sees such material arrangements as being made up of entities of four types—"human, artefacts, organisms and things of nature" (ibid., 129), interconnected as arrangements but to some degree having distinct characteristics and roles in social life, and their own forms of interaction: "direct relations among material entities … contribute to the shape and progress of social life" (ibid., 135). While the term 'things' acts as a cognitive metaphor for discrete objects, by focussing on 'actions' and 'direct relations' Schatzki enables us to engage with how matter is in-relation. One key form of relation is energetic. Heat is one mode through which energy is expressed, and it can be accounted for through thermodynamics, thermal physiology, or more broadly as thermal flow.

Schatzki comes briefly to a more specific engagement with such flows in a working through of the different ways in which materiality is part of society. He explains that "biological and physical flows pass through practice-arrangement nexuses" and indeed that material arrangements are "in some sense crystallisations of matter-energy flows" with such flows "also mediated by practices" (ibid., 137). He is clear that materiality in its various forms can precede and exceed the social and should not be assumed to be embedded within practice configurations; "matter-energy flow[s] through practice- arrangement nexuses regardless of whether and how labour and technology channel, shape or capture the flows involved" (Schatzki 2010, 138). There is a need to be careful about the interpretation of the word 'regardless' here; flows per se may be inevitable, but the nature of those flows can indeed be shaped by labour and technology. At the same time, labour and technology are shaped by these matter-energy materialities, while other flows may never or rarely come into contact with social practice-arrangement nexuses. Schatzki (2010) therefore provides a way of reconciling the extensiveness of energetic-material flows with the particularities of how heat figures within the configuration and enactment of certain social practices.

Two key insights emerge from this way of conceptualising the immersion of social practices in thermal flows. The first is that, in its flowing-through all practice-material arrangements, heat is productive of practice performances. 'Productive' here is meant in the sense that thermodynamic flows are part of the necessarily present material arrangements that (all) practices are reliant on. All social practices (as conventionally understood) involve human bodies that, as already outlined, have a thermal 'operating range' within which they are able to readily function. For other practice entities, there are also shared understandings of the thermal conditions required for their successful enactment (such as yoga 'needing' a warm room). Many technological artefacts that are enrolled as elements in practice configurations also have thermal operating ranges—beyond which they may be unable to work effectively (Shove et al. 2014). The dynamic heat 'environment' within which they operate is therefore productive of practice by enabling these elements, and thus the practices they are part of, to operate successfully.

It follows that if thermodynamic balances (as embedded within bodies, spaces, or technologies of practice performance) do *not* remain within normal or necessary ranges, then the flow dynamics required to keep all of these elements operational will be less able to support successful practice performances. Indeed, harm and damage may result from these performances instead. As Walker (2013) argues, the literature on practices has been more concerned with distilling what makes practice performances successful rather than what explains their failure or problematic outcomes. Thermal flows are a case in point; while heat is essential to all practice performances, problems occur if the volume or rate of these flows changes, such as when an engine 'overheats' as part of a practice performance of driving.

This leads to the second insight about the immersion of practices in thermal flows: heat is used in, but is simultaneously also a *product* of practices. In the course of performing practices, heat is produced, dispersed, or stored, meaning that practices are intervening in various ways in thermal flows that are running through them. Jerstad (2016) provides a detailed account of how vigorous work practices warm bodies in cold weather, such that the body's metabolism becomes part of the 'energy mix' that manages the seasonal absence of environmental heat. Shove et al. (2014) also detail how technologies integral to practice performances (such as computers) can produce waste heat such that further intervention is needed to cool down the technology and the spaces it inhabits.

These forms of heat generation can be readily integrated into shared understandings of how particular practices are configured and enacted, but the *point* at which they do so, and *how* this is possible, is shaped by extant energetic-flows and energetic-material relations. As such, whilst heat is both productive and a product of social practices, these dynamics are closely intertwined. Walker et al. (2014) demonstrate how such a thermal balance, integral to practice performances, can become destabilised in an example of technology in an intensive care unit. Here the heat production of the unit's machinery, while essential to its operation, nonetheless undermined its ability to continue functioning once the multiplication of units reached an imbalance with the thermodynamics of the room that housed them; heat-as-product undermined heat-as-productive of practice performances.

Such material-energetic dynamics emerge temporally and spatially in ways that are under-analysed in many accounts of practice. Shove et al. (2014) draw on Ingold's (2007) notion of living as 'mingling' with the weather to engage with this issue, but gloss over the contingencies through which this occurs. Ingold, while making 'visible' pervasive yet occluded energetic-material relations, nonetheless still occludes the particularities of energetic-material mingling (its forms, speeds, dependencies) through a generic account of flows. To take Walker et al.'s (2014) example of the intensive care unit, in a fixed-size room with finite pieces of equipment, the potential heat load might seem relatively straightforward. However, the amount of work those machines do, and therefore how much heat they generate, is in dynamic relation with the number of patients, illness types, and severity of condition that the machines manage. This changes from day to day and hour to hour. As such, the components and pattern of distribution of materials and energies matter, but so too do the rates and intensities of material-energetic flows, understood in this instance through thermodynamics.

Weathering Work Practices in Northern Australia

To exemplify the significance of thermal flows in practices, this section draws on a mixed methods study of the practices of critical infrastructure workers in Darwin, northern Australia, in 2015. Operating close to the edge of the thermal envelope that human bodies can tolerate, the role of thermal energy as both productive of practices and produced by them becomes much more perceptible, particularly in terms of the negative effects of excess bodily heat. The study demonstrated how successful practices depended on the appropriate management of the material-energetic economy of heat. In part, this management was enabled by practitioners' ability to 'know' excesses and deficits of thermal flows somatically and experientially. Excerpts from ten semi-structured interviews with crew and their managers (identified using pseudonyms) are used below to present workers' reflections on the imbrications of heat and the practices of work.

Situated Relations Between Thermodynamics and Thermal Physiology

Northern Australia's tropical monsoon zone combines high levels of heat and humidity. Such environmental conditions are particularly taxing for the human body, as its thermoregulation system struggles to dissipate unwanted excess heat through exchanging it with the surrounding air and surfaces (Oppermann et al. 2017). Here thermodynamics and human thermal physiology are intimately related; high environmental temperatures reduce the thermal gradient between the body and the air, so heat does not easily travel by convection from one to the other. The saturation of the air with environmental humidity also curtails the evaporation of sweat, such that heat cannot be easily transferred away from the body (Mitchell et al. 1976). As such, the thermal flows in which practices occur become noticeable through the over-accumulation of heat in the bodies of practitioners, sensorially apparent through thermoception, and indirectly as fatigue, nausea, or cramps, and the associated production of excessive sweat and thirst.

As a place, tropical Australia is often characterised in terms of the somatic experience of energetic excess: it is known for 'mango madness' where populations 'go troppo' as the physical over-accumulation of heat in the body plays havoc with cognitive functioning manifesting in anti-social, erratic, and violent behaviour (Oppermann et al. 2017). The usually invisible dependence of 'social,' 'regular,' or 'peaceful' behaviour on appropriate stocks and flows of thermal energy is thrown into relief here, as is the intimate, yet largely ignored thermal constitution of everyday practices that assume these capacities.

The workers in the study were responsible for building and repairing electricity transmission lines outdoors. Immersed in intensive thermal flows on a daily basis, managing these flows was integral to their ability to complete their work. The regulation of the accumulation/loss of heat took several forms. The most obvious was managing the temporality of exposure to hot environmental conditions. As one job scheduler noted, "we always factor in a bit of extra time [...] and ensure we have an extra guy or two if we can, so that we can rotate through." Similarly, work at night-time was valued as an alternative mode of managing exposure to environmental heat—here, rather than rotating staff within a job, it is the rotation of the earth that enables regular relief from the constant 'topping up' of bodily heat from solar thermal flows.

In addition to reducing exposure, the rotation of staff is also effective because it manages exertion. This regulates thermal flows by limiting the heat the body itself produces through physical work. Whether hauling steel cables or simply breathing while sleeping, the exertion-related endogenous production of metabolic heat is an inescapable part of all social practices. However, it becomes particularly noticeable in already-hot environments where there are high levels of exertion as heat accumulates much faster than it can be dissipated, taking the body to higher than comfortable (or safe) temperatures. In the organisation where the workers were based, this was redressed through incorporating periods of rest or rotations to less active roles, not only within but also across jobs; "...we try to avoid a lot of back-to-back big jobs. So, if we're going to have a big job today, we'll have something that's not quite as taxing the next day."

These environmental thermodynamics and their imbrications with bodily thermoregulation and endogenous heat production mean that work practices are infused with complex material-energetic dynamics. Key to enabling the successful regulation of *how* and *where* heat is produced, absorbed, or lost are the *durations* and *frequencies* with which performances of work practices occur in sequences and temporal patterns. These temporalities will be explored in more detail in the next section.

The Temporalities of Heat in Practice Performances

Each performance of the practice of critical infrastructure maintenance also produces a temporal patterning in the accumulation and loss of bodily of heat. In the study, these temporalities could be found in durations of how long work continued; the pace, intensity, and amount of work completed; and the rhythmic effects of sequencing and repetition of changes in exposure or exertion. As such, the heat which workers absorbed, generated, and lost was intimately related to the rhythms of 'doing' repairs and vice versa. This temporal patterning of body heat storage was thus produced by the assemblage of practice elements of meaning, competences, and materials and—crucially—how they manage and produce thermal flows in practice performances over time.

For example, the materialities of the location and environmental conditions of the repair site and the meanings attached to the work (such as its urgency, complexity, or importance) shaped the frequency of hydration and active cooling, as well as whether these were enrolled in or seen as disrupting the activity of work. Hydration is closely linked to thermoregulation's ability to enable heat loss, whilst dehydration is a product of the body's attempt to regulate temperature through sweating. In the study, the physical logistics, and perceived normative value, of hydrating while performing work above ground level were clearly observable during the site-based ethnography and were also apparent in the reflections on heat management by the crew:

> *[being on the ground means] it can be easier to get to your water. You don't have to yell out to the person on the ground [to refill and throw the bottle up] or come down and grab it when it's empty. [...] If you're busy doing something, you run out of water, you're probably going to finish doing it if you're in the air and then come down. (George)*

Such micro-decisions, conducted throughout the course of the day, shaped cumulative levels of bodily heat and (de)hydration, directly affecting workers' physical capacity to continue work. However, explicit decisions to manage heat stress, such as by managing hydration levels, were only one aspect of work-heat temporal relationships. Sometimes the way in which work was scheduled meant incidental, but no less valuable, periods of cooling were part of the daily pattern of work practice. As George noted: "... even if it's just driving to the next job you're going to be in the air conditioner. You're cooled down."

Thermal flows thus pervade all aspects of sequences of practice performances. Balancing bodily heat accumulation and loss is not only a constant endeavour, it also has complex social and physiological causality with time lags of a few minutes to a few days. In already extreme environmental conditions with their intensive flows of thermal energy, George's account of his decision process highlighted how fine-grained such temporal considerations of heat were in managing hydration:

> *...if it's ... something quick, you'll finish it off... but only if it's a few minutes. If you're going to be up [a power pole] for another 10 minutes and you're out of water, you're just going to [have to] come down and fill [the bottle] up. (George)*

Sometimes awareness of accumulated bodily heat and anticipation of its effects were hard to assess and manage, as there could be a considerable time lag. Recalling an incident of a colleague suffering from heat illness, Zac noted that despite heat being accumulated through work, "...he didn't get the dizzy spell until he was driving home."

The temporality of thermal flows is absolutely critical to the health of practitioners' bodies. In the examples above, practitioners respond to actively harness and inadvertently shape the temporalities of thermodynamics as they play out across matter and space. The amount of work over time and its intensity also produce a *rate* of work that generates varying amounts of endogenous heat that 'matters' energetically because of its effects on the stock of heat in the body. In turn, this affects other bodily processes which require certain temperature parameters to function effectively. Excessive stocks of heat also produce material damage in tissues and cells and at the sub-cellular level. Here too, time is a factor. In fact, thermal damage to tissues is measured not just as temperature but as temperature *over time*, as cumulative equivalent minutes at 43 °C (Yarmolenko et al. 2011).

In thinking about how practices are constituted and performed then, we need to pay attention to heat (and by extension, all energy) in terms of its dynamism—as both change and temporality (Walker 2014). Energy does not sit still. It moves and degrades. If it 'stops,' it needs to be expended in material effects or stored as fuel. To move again, it must be converted and released. Energy's dynamism means it is inherently temporal, but not necessarily linear, as the materials through which it passes (material-energetic relations) shape the rate at which energy moves and the forms it takes. In practice terms, the dynamics of energetic flows are inherently related to the practice elements through which they move: not only material elements of practices, but the way these are connected, distributed, or utilised by the skills and meanings that are also practice elements. As such, dynamic matter-energy relations constitute practices and are productive of practices' temporal patternings. At the same time, the way particular practices are constituted can also shape matter-energy relations and hence energetic flows.

Observing the requirement to manage the economy of heat in the human body within the wider economy of heat decentres the human in social practice theory by showing its inherent dependence on and co-production with material-energetic flows. It also embeds what it is to be human in its very material nature, demonstrating how bodies and practices, intentionally and unintentionally, actively modulate thermal flows and material-energetic relations down to a cellular level.

The Visibility of Thermal Flows in Social Practices

Rendering thermal flows 'visible' in everyday life is difficult, as the situated performance of practices can be finely attuned to their thermal environment. As Ong (2012, 16) notes: "If we stop to consider [it,] our daily routine is punctuated by what are essentially acts of homeostasis—adjusting the temperature of our baths […]food and drinks…." As such, it is not necessarily the case that very high or very low temperatures will disrupt what are understood as 'normal' performances of practices. Rather, "the qualities of an environment can be considered in relation only to specific forms of life" (Ingold 2007, S26, referring to Gibson 1979). For example, to Inuit living in some of the lowest temperatures to harbour human life, these temperatures are not 'extreme' but rather inherent to established practices of everyday life. In fact, the impacts of climate change in warming these low temperatures to what others might consider more 'hospitable' levels can produce severe problems for the historically established practices that constitute local patterns of Inuit living (Vannini et al. 2009). As such, excess and deficit cannot be defined by a universal abstract norm or thermal ideal, but rather by the degree to which thermodynamic patterns are successfully enrolled in practices for human life—or perhaps more accurately—the degree to which social practices successfully enrol human life in thermal patterns.

The material-energetic dynamics of heat in practices thus become clearer when *disruptions* of these co-productions occur, rendering their constitutive components (elements and flows) more visible. For this reason, in the study the operations and temporalities of the thermal flows through material-energetic relations were most observable when practices of heat management 'failed' (even though practices of work might have succeeded) or when practices of work 'failed' because the impost of heat became too unbearable. In the face of such failure (anticipated or experienced), the circulations of heat which social practices both rely on and regulate may have to be so actively managed as to take on a very overt, even determining role in shaping practice performances (Jerstad 2016).

Excesses or deficits of heat then are defined in relation to practices, themselves developed in relation to environmental (such as climatic and seasonal) patterns. Sometimes, thermal flows shift because practices as a whole relocate, at which point their dependence on implicit environmental conditions can become more obvious. An example of this is the way that colonial conquest moved European practices from temperate regions to tropical regions. The characterisation of environmental heat as 'extreme' in the northern Australian case study can be seen in part as an effect of such a translocation of practices, as European norms of working hours, protective clothing, and performance management assume a very different climate to that in which they are now taking place. Yet the study also demonstrated that constant 'exposure' or 'immersion' resulted in certain shifts in these work practices; either the homeostatic effects of the practice was exceeded in a particular performance which was then modulated to accommodate its new thermal flows, or, if performances were routinely exceeded, the practice as entity could change, demonstrating the productive effects that thermal flows can have on, and in, social practices.

In the context of such profound physicality, the ways in which thermal flows, and material-energetic relations more broadly, are *known* come into question. Somatic and sensory awareness of heat is certainly possible. In the study, the more experienced workers felt they had developed an enhanced bodily awareness of both heat and heat stress and as a result also had the ability to actively utilise this awareness to practice work in a more responsive fashion. In urgent repairs, workers used their bodily awareness of how much heat they could handle to push their bodies to work longer or harder. On the other hand, in low urgency repairs, they might more actively manage heat exposure and production through resting, seeking shade or air-conditioning, or ingesting ice to prevent or respond to symptoms of heat stress (such as feeling thirsty, having a headache, losing clarity of thought) more quickly. The competence of being able to listen to and assess the status of one's body, project heat accumulation and its impact on capacity to work, and adaptively pre-empt, prepare for, or recover from heat impacts is a highly complex form of modulation of thermal flows, played out at an everyday scale.

In the practices of critical infrastructure work in extreme conditions, such sensory awareness and ability to manage how much heat they accumulate is essential to practitioners being able to work successfully. To play on Ingold's (2015) phrasing, such competence enables practitioners to 'weather work' but also to 'work weather.' The somatic embeddedness and sensory awareness inform a sensory competence around how to 'do' work, including the deliberate tactical management of multiple material-energetic patterns and rhythms, in constant and contingent relationship with one another. In the study, there were a plethora of thermal relations, gradients, and trajectories to consider, including the passage of the sun; changing temperatures or humidity; repair types, schedules, and crew numbers; exertion, exposure, and rest; logistics and equipment; meal times; and opportunities for passive or active cooling and hydration. In this turbulent, constantly moving medium of matter-energy relations, heat was sometimes perceptible and sometimes imperceptible, sometimes explicit and sometimes implicit, but always inescapably present—productive of practices and produced by them, both as practice-as-performance and practice-as-entity.

Discussion and Conclusion

We have considered the status of heat as an (im)perceptible dynamic non-human that is both productive of practices and produced by practices. We have argued that all social practices—deliberately or indirectly—draw on and intervene in thermal flows: hastening, retarding, redirecting, collecting and distributing, converting, or producing thermal energy. Humans, animals, machines, and plants all have thermal ranges or equilibriums, which they must maintain to function well. The dynamic nature of thermal energy means that maintaining any system at an ideal level is a constant practice. Across much of the plenum of social life, these dynamics of heat flow silently through how social practices are understood, performed, and reproduced, always present but not necessarily overtly agentive. When we look carefully, we see homeostasis actively being produced in relation to excesses or deficits, yet these dynamics too are part of social practices themselves.

Understanding everyday life in terms of practices has helped scholars break it up in ways that make it more open to analysis and intervention. We have drawn on Schatzki's work to reconsider how heat can be understood in terms of social practices. Heat here is understood not as 'an' element of practices, but as elemental, flowing through practices. Even as matter-energy relations shape thermal flows, humans also act on these flows through a series of corporeal and non-corporeal interventions, slowing, hastening, or generating thermal energy in and out of their own bodies or other materials and objects. Thermal flows, then, are one register through which to explore notions of human/non-human hybridity that Wallenborn (2013, 152) describes as 'extended bodies' constituted "through flows including those of air, light, … energy." Understanding practices in terms of their implicit as well as explicit utilisation and production of heat, as well as other energetic-material relations, helps to decentre the human in our account of a social practice, noting their contingent emergence in the extensive flows that far exceed human knowledge or control. This helps make visible the more-than-human energetic register and energetic-material relations through which social practices occur. It also provides a richer sense of their contingency, inherent dynamism, and potential to change.

In relation to heat and its role in practices, the 'location' of transformation is not only in changing elements within practices but in the nature and temporality of energetic flows on which they depend and which occur within and between practice elements and within and between practice performances. The key then is not to see heat as an 'object' or 'outcome' to be controlled, but rather as a process that is continually influenced or modulated. Nor are humans capable of completely managing thermal flows at any scale, from the bodily to the planetary. This paradox of being at once profoundly embodied, tangible and material, and yet at the same time profoundly vulnerable, contingent and exposed has a somatic immediacy in the bodies of those most exposed or exerting themselves in hot environments, particularly as those energetic flows become more intensive as a result of the heat stored by global greenhouse gases.

Drawing these theoretical reflections together with our empirical study of everyday practices, we have seen that outdoor, labour-intensive work in extreme conditions is a constant, tactical process of managing heat accumulation and loss. This produces patterns of activity that limit exposure or allow for recovery with sufficient amounts of time to ensure heat does not reach (or remain at) dangerous levels in the bodies of workers. As such, the heat that workers absorbed, generated, and lost was intimately related to the rhythms of 'doing' repairs; and the rhythm of the repair was intimately related to the temporal patterning of bodily heat storage and loss.

The imbrications of bodies in thermal flows point to the need to consider the material-energetic flows that are productive of, and produced (or at least, shaped) by, social practices. Schatzki (2010) provides an ontology of social practices that is open to such dynamic relations, and, as we have demonstrated, this provides a starting point for finding ways to acknowledge and account for energetic flows in all of their dynamic changeability and temporality. There are, of course, many applications for this—thinking about the role of calorific energy in practices of food production, eating and exercise, or sound energy in practices of construction, design, and travel. Such an endeavour is an ongoing task, requiring diverse modes of attention and extending beyond a predisposition to 'things' and 'objects' that are visible, discrete and mechanically connected, to an attentiveness to flows and contingency. We should also be cautious not to fall back on uncritical epistemologies and methods in seeking to make visible energetic-material relations. In the case of heat, thermodynamics and thermal physiology are very useful and productive, but so too are somatic, embodied, and experiential ways of knowing the world, as demonstrated in the accounts of outdoor manual workers. Acknowledging and cultivating more-than-human materiality, not only as a meaning of practices but as sensory skill or competence, may be one of the ways in which social practice approaches can help engage with the energetic register of everyday life, *for* everyday life. In the particular case of heat, social practice theory can support this by making the invisible visible through acknowledging and demonstrating the dynamic, elemental, and sensorial ways in which heat produces, and is shaped by, practice performances.

References

Bataille, G. 1988. *The Accursed Share: An Essay on General Economy*. Translated by Robert Hurley. Vol. 1: Consumption. New York: Zone.

Clark, N. Forthcoming. A Short History of Combustion. In *What is Energy and How We Might Think About it*, ed. S. Kwinter and K. Moe. New York: Actar, 2017.

Hui, Allison, Theodore Schatzki, and Elizabeth Shove, eds. 2017. *The Nexus of Practices: Connections, Constellations and Practitioners*. Abingdon: Routledge.

Ingold, Tim. 2007. Earth, Sky, Wind, and Weather. *Journal of the Royal Anthropological Institute* 13: S19–S38. https://doi.org/10.1111/j.1467-9655.2007.00401.x.

———. 2015. *The Life of Lines*. Abingdon: Routledge.

Jablonski, Nina G. 2013. *Skin: A Natural History*. Berkeley: University of California Press.

Jerstad, Heid. 2016. Mundane Energies: The Working Body as a Heat Source in the Indian Himalayas. *Anthropology Today* 32 (4): 7–10.

Maller, Cecily. 2018. Dynamic Bodies in Theories of Social Practice: Vibrant Materials and More-than-Human Assemblages. In *Social Practices and Dynamic Non-humans: Nature, Materials and Technologies*, ed. Cecily Maller and Yolande Strengers, 87–107. Cham, Switzerland: Palgrave Macmillan.

Mitchell, D., L.C. Senay, C.H. Wyndham, A.J. van Rensberg, G.G. Rogers, and N.B. Strydom. 1976. Acclimatization in a Hot, Humid Environment: Energy Exchange, Body Temperature, and Sweating. *Journal of Applied Physiology* 40 (5): 768–778.

Ong, Boon Lay. 2012. Warming Up to Heat. *The Senses and Society* 7 (1): 5–21. https://doi.org/10.2752/174589312X13173255801969.

Oppermann, Elspeth, Matt Brearley, Lisa Law, James A. Smith, Alan Clough, and Kerstin Zander. 2017. Heat, Health, and Humidity in Australia's Monsoon Tropics: A Critical Review of the Problematization of 'Heat' in a Changing Climate. *Wiley Interdisciplinary Reviews: Climate Change*: e468–n/a. https://doi.org/10.1002/wcc.468.

Reckwitz, A. 2002. Toward a Theory of Social Practices: A Development in Culturalist Theorizing. *European Journal of Social Theory* 5 (2): 243–263.

Rinkinen, Jenny. 2018. Seeing Wood for the Trees: Placing Biological Processes Within Practices of Heating and Harvesting. In *Social Practices and Dynamic Non-humans: Nature, Materials and Technologies*, ed. Cecily Maller and Yolande Strengers, 47–62. Cham, Switzerland: Palgrave Macmillan.

Schatzki, Theodore. 2010. Materiality and Social Life. *Nature and Culture* 5 (2): 123–149.

Shove, Elizabeth, Mika Pantzar, and Matt Watson. 2012. *The Dynamics of Social Practice: Everyday Life and How It Changes*. London: Sage.

Shove, Elizabeth, and Nicola Spurling. Ed. 2013. *Sustainable Practices: Social Theory and Climate Change*. Abingdon: Routledge.

Shove, Elizabeth, Gordon Walker, and Sam Brown. 2014. Material Culture, Room Temperature and the Social Organisation of Thermal Energy. *Journal of Material Culture* 19 (2): 113–124.

Vannini, Phillip, Godfrey Baldacchino, Lorraine Guay, Stephen A. Royle, and Philip E. Steinberg. 2009. Recontinentalizing Canada: Arctic Ice's Liquid Modernity and the Imagining of a Canadian Archipelago. *Island Studies Journal* 4 (2): 121–138.

Walker, Gordon. 2013. Inequality, Sustainability and Capability: Locating Justice in Social Practice. In *Sustainable Practices: Social Theory and Climate Change*, ed. Elizabeth Shove and N. Spurling. Abingdon: Routledge.

Walker, Gordon. 2014. The Dynamics of Energy Demand: Change, Rhythm and Synchronicity. *Energy Research & Social Science* 1 (1): 49–55.

Walker, Gordon, Elizabeth Shove, and Sam Brown. 2014. How Does Air Conditioning Become 'Needed'? A Case Study of Routes, Rationales and Dynamics. *Energy Research & Social Science* 4: 1–9.

Wallenborn, Grégoire. 2013. Extended Bodies and the Geometry of Practices. In *Sustainable Practices: Social Theory and Climate Change*, ed. Elizabeth Shove and N. Spurling. Abingdon: Routledge.

Yarmolenko, Pavel S., Eui Jung Moon, Chelsea Landon, Ashley Manzoor, Daryl W. Hochman, Benjamin L. Viglianti, and Mark W. Dewhirst. 2011. Thresholds for Thermal Damage to Normal Tissues: An Update. *International Journal of Hyperthermia* 27 (4): 320–343.

Part II

Technologies, Automation and Performativity

8

Displacement: Attending to the Role of Things in Theories of Practice Through Design Research

Ron Wakkary, Sabrina Hauser, and Doenja Oogjes

Introduction

In this chapter, we attend to the roles of non-human material elements or what we refer to as *things* in everyday domestic practices. We discuss our recent interaction design research that aims to move beyond a human-centred view of digital artefacts in the home to reveal a less resolved and more complex picture of things, and the inherent challenges in understanding things beyond human perspectives. Having relied on theories of practice (ToP) in the past to inform our work, we see value in

R. Wakkary (✉)
School of Interactive Arts & Technology, Simon Fraser University, Surrey, BC, Canada

Eindhoven University of Technology, Eindhoven, The Netherlands
e-mail: rwakkary@sfu.ca

S. Hauser • D. Oogjes
School of Interactive Arts & Technology, Simon Fraser University, Surrey, BC, Canada
e-mail: shauser@sfu.ca; doogjes@sfu.ca

C. Maller, Y. Strengers (eds.), *Social Practices and Dynamic Non-Humans*,
https://doi.org/10.1007/978-3-319-92189-1_8

further discussion about how *things* are understood in ToP. Our work reveals that to do so seriously one runs headlong into the non-human nature of things and their inescapable opacity and instability. As a result, and unsurprisingly, things withdraw from human understanding and make themselves less than transparent to ToP or interaction design. In this work, we aim to show that in tracing practices through things, the relations between things and practices are not so readily resolvable. Further, the integration of things in practices may only emerge in an incomplete picture of indirect traces and residues of non-human elements or indirect human engagement.

As interaction design researchers, our research has focused on the home as a site of inquiry, and it is the things of domestic practices, ranging from chairs or tableware or other everyday things to novel digital artefacts, that we ourselves craft and deploy as objects of study. It may seem obvious that as designers, we focus on things, yet in our fields of interaction design or human-computer interaction (HCI), researching and understanding things is obscured or distorted by an overwhelming focus on human matters. Human use and perception of technologies are long-held and privileged assumptions for how to understand and frame design and technologies. This deeply human-centred view is manifest in HCI researchers' focus on technology as a prism through which to view human behaviour, human perceptions, and human social relations. Design is governed by the principles and assumptions of human-centred design in which human concerns hold primacy in the designing of digital artefacts. As a result, our knowledge of things is bound up in an unreliable asymmetry of—or overshadowed by—human concerns and perceptions. This leads to a limited understanding of the role of things in our world as matters of human functionality and perception. We believe that within HCI a clearer and deeper view of the relations of things and humans will productively decentre the focus on individuals and individual human behaviour and lead to a framing that allows for more complex but integrated units of study, such as human-technology hybrids or practices as a hybrid unit of meaning, materials, and competences.

In this context, our investigations on and through things are aimed at revealing the possibilities to rethink relations between things and us.

In our work, we move beyond a human-centred understanding of things by embracing its hybridity with humans, that is, how things and humans co-constitute or mutually shape each other. In this regard, we do not abandon human experiences to solely pursue the experience of things, in large part because the experiences of being a thing are not fully available to us as humans. It may seem obvious, but it is surprisingly overlooked that any understanding of non-human elements poses inherent intelligibility issues, in that any understanding is limited or at least mediated by human perspectives, perceptions, and reasoning. Hence while we desire to reveal subtleties of the relations of things to people and also things to things, from an empirical standpoint, a complete picture of the dynamics of non-human elements in practice is an impossibility to describe. More to the point what is made intelligible to us is often elusory and indirect, as we will examine further in this chapter.

Our aim is then to pursue a thing orientation as a way to bolster our understanding of the hybridity between us and things while operating within the conditions of hybrid relations. In short, this means that we accept the bounded and mutual shaping of relations of things and us without fully knowing the particularities of the hybrid relation and how things take on their role within it. We extend this hybrid framing of things and humans by weaving our work with the theoretical notions of postphenomenology (Rosenberger and Verbeek 2015) and social practices, building on the need to understand the hybridity of human and non-human elements in practice (Shove et al. 2007) and the true status of material elements in ToP (Reckwitz 2002). As a result, we do not question that non-human elements are fundamental to social practices but rather that the particularities of how things play that role are more complex and elusive than currently understood. The challenge of understanding things in ToP is equally fraught with intelligibility issues, and in part, ToP may suffer the same human-centred imbalances as HCI, thus obscuring or resolving too quickly the role of things in practices.

The crossing of ToP and interaction design research is mutually beneficial. Interaction design is concerned with the interactions between technologies and people. ToP offers a reconceptualisation of interaction design whereby practice becomes the object of action, inquiry, and concern rather than individual use of technologies. While ToP can shed light on and constructively problematise a fundamental notion like *use* in interaction design, interaction design in turn offers material and empirical understandings of the particulars of things or non-human elements that shed light on the nuances, contradictions, and problematics that get in the way of a ready conceptualisation of things in practice.

There are many instances in ToP where things like paint for the DIY painter (Shove et al. 2007) or walking poles for the Nordic walker (Shove et al. 2012) are easily resolved in identifiable practices through function and meaning. However, does this approach account for all things in practices? What about things that are not so readily understood within a given practice? The things we later discuss in this chapter can be seen as related to what we loosely refer to as domestic practices as part of an empirical and speculative inquiry, which we will explain in detail further on in the chapter. This discussion evokes the myriad of practices related to everyday living within a home from cooking to entertaining to doing the laundry. However, we do not specify which of these practices if any are in relation to the things we have designed. In fact, the specifics of any given practice fade into the background, and so while practices are not specified, they remain present and do not disappear. In our analyses, things come to our attention in the 'loose-knit patterns' of bundles of practices as Shove and colleagues describe (Shove et al. 2012). We assume the ongoing engagement, in any form, over time affirms that things are indeed integrated into some practices but exactly in what way and in which practice we are not so quick to resolve or identify. Herein the difficulties with the intelligibility of things make their presence clear. In our research, it can be said that we purposely elicit this type of ambiguity as a means of investigation. We aim to show that tracing practices through things offers a cloudier view of what is being performed or practised. Similarly, through exploring the practices of a robovac in Chapter 11, Strengers challenges human intelligibility, sensibility, and practical accomplishment as concepts in social practice theory. More than not, the sayings and doings of things withdraw from sight and sound and are evident only in the indirect traces and residues of non-human elements or indirect human engagement. However, we assume that things are integrated into practices with an ongoing effect on everyday life even if we cannot identify the practice. This difficulty and our assumptions reveal, in relation to practices, the need to develop concepts that help account for the traces of things rather than the things themselves.

In this chapter, we offer our design theory concept of *displacement*, which we developed and utilised in our research to navigate the challenges of tracing the relations of things and, in this case, things in relation to practices. We trace how displacement was informed by conceptualisations from philosophy of technology. In particular, we point to the condition of *withdrawal* in which there are limits to our understanding of things given that the experience of things recedes from our ability to perceive and understand (Bogost 2012; Verbeek 2005). Relatedly, we also describe *multistability*, the variant perceptions of things relative to different embodied perspectives and actions (Ihde 1990). Within this theoretical context, we utilise displacement as a way to frame the partial and shifting visibility of relations with things or the role of things in practices as a matter of residues, traces, or displaced effects of human or non-human actions.

We ground this chapter in a theoretical discussion highlighting how issues of relations of things emerge from our design research, namely, our investigation into relations of things and us, informed by ToP and postphenomenology. Thereafter, we attend to our main concept of displacement, which gives insight into the role of things in practices from a non-human-centred perspective. We focus on explicating the concept and related ideas through two research artefacts we designed and deployed known as *Morse Things* and *table-non-table*. We conclude the chapter by reflecting on how this approach can help articulate the role of things in social practices that can lead to deeper, less human-centric understandings.

Theoretical Understanding of Hybridity and Non-human Orientation to Things

A theoretical contribution of this chapter is the weaving together of our design research that is grounded in our previous work and philosophical perspectives on things with ToP. What follows is an overview of our investigations into everyday design and a discussion of relevant philosophical insights into things and human-thing relations. Throughout this section we tie our work to perspectives and theories of practice.

Our path to an investigation of thing relations in interaction design began with the development of our concept of *everyday design*. This concept emerged from our studies of various everyday practices such as family life (Wakkary and Maestri 2008), sustainability-oriented practices (Wakkary et al. 2013), and expert amateurs (Desjardins and Wakkary 2013). In essence, this research argues that every person is a designer and that everyday design describes an ongoing process of resourcefully using and appropriating everyday artefacts as a means to construct everyday life on an ongoing basis. For example, we have observed the unique assembly of a chalkboard, a hanging fruit basket, the fridge door, and sticky notes to create a complex but effectively situated way to relay messages to family members (Wakkary and Maestri 2008). As a matter of clarity and motivation, it is important to note that ToP, especially performances of practice (Schatzki 1996) and the elements from which practices are made (Shove et al. 2012), informed ongoing refinements of our concept of everyday design. We incorporated perspectives on practices into our work, and it served as a bridge between our earlier descriptive ethnographic research and our more recent generative design research.

The implication of everyday design for interaction design is that there is an ongoing shaping of the use, meaning, context, and value of artefacts by humans or what we refer to as everyday designers. The meaning, value, and identity of both the person and the artefact change through the interaction representing the ongoing configuration of one's everyday life. We have over time arrived at an understanding of this as a type of co-shaping or mutual agency between artefacts and people through postphenomenological perspectives. Postphenomenology, a strand of philosophy of technology, understands technology or things through the relations human beings have with them (Rosenberger and Verbeek 2015). Technologies are seen as mediators of human existence and experiences rather than separated functional and instrumental objects or alienating entities (Verbeek 2005). Humans and technological artefacts mutually shape each other through mediations that form human subjectivity and objectivity of any given situation. A consequence of this mutual agency is the inevitability that artefacts themselves enter and shape practices through the invitation or inhibition of particular aspects of actions or through the amplification or reduction of aspects of experiences (Verbeek 2005). In this sense, agency whether of humans or things is not unfettered. Rather than viewing humans or things as autonomous, informed by the postphenomenological perspective, we see them sharing a relational form of agency that hybridises humans with things and vice versa but still allows for making a distinction between them. Both postphenomenologists and sociologist Bruno Latour, whose work has informed several practice theorists, talk about this hybridity but approach it in different ways. Latour sees these hybrids as finding no place in a subject-object schema (Latour 1993), whereas in postphenomenology, '[h]ybrids arise in the form of complexes of humans and technologies' (Verbeek 2005, 162), which are considered distinct but intertwined through mutual constitution or mediation. This is where a postphenomenological understanding of mutual agency differs from Latour's. In Latour's perspective, agency is not seen as delegated and distributed to human and non-human actors, but rather is relationally constructed through practice (Latour 1993). We assume a position closer to postphenomenology that allows for distinctions to be made between humans and things while considering them as a hybrid. As a result of our migration towards this philosophical perspective, the emphasis of our design research shifted to exploring the relations between people and things rather than the human behaviour or qualities of things in themselves.

Within ToP, most scholars recognise a need to acknowledge the importance of things in practices based on Latour's position in which humans are not the only participants in practice but that non-human artefacts are seen as equally important actors (Latour 1987). For example, Reckwitz (2002) invokes his attention to things, as well as Shove and colleagues (Shove et al. 2007) who see the importance of understanding objects and materials with respect to theories of social practice and who explore the distributed competence between the object (tool) and humans by drawing on Latour's ideas that competence is divided between the embodied knowledge (in the human) and the embedded knowledge (in the object and materials) (Latour 1987). However, Shove et al. (2012, 22) also see the distribution of competences between people and things as a complication of Latour's human-non-human hybrid to include diverse human and non-human entities. Reminiscent of postphenomenology, Shove and colleagues argue for the need to 'focus on the co-evolution of these hybrid entities rather than on the human or non-human elements alone' (Shove et al. 2007, 59). In this chapter, we do indeed focus on the things or material elements but as a matter of tracing the hybridity between the diverse human and non-human entities.

In summary, in this section we have discussed how we have threaded together our understandings of ToP and postphenomenology to scaffold our non-human-centred approach to the role of things and to refine and articulate the hybrid relation of things to other material elements of practice. Next, we discuss how this theoretical grounding has informed our design research in what can best be described as empirically speculative investigations into the role of things in domestic practices or what we refer to as material speculations. This design research has led us to the philosophical notions grounding our approach and the concept of displacement.

Material Speculations

As we stated earlier, ToP and interaction design can fruitfully cross paths. While ToP acknowledge the importance of materiality, this body of theory arguably foregrounds people and therefore focuses on the practices of humans and the material elements that are in direct relation to humans and human perception. For example, when the practice of showering is studied, the material elements attended to include water, soap, and the shower head. Yet, in relation to this configuration of material elements of showering, there are many other elements that are left unstated or unobserved such as condensation on a mirror, mould on the bathroom wall, decorative porcelain tiles, batteries from a Bluetooth speaker, or a ceramic ashtray. Within the confines of the human practice of personal hygiene, these are obviously filtered out as co-located but unrelated elements to the practice. But how can we make sense of these things? This is not to say that these things are parts of 'arrangements' as Schatzki (1996) would argue, describing them as features of the setting or contexts separate from practices. We rather view these things as also constituted within practices, however unclear or unknown these practices may appear. As we stated earlier, we aim to show that in tracing practices through things a cloudy view emerges in which things appear displaced and decentred. This can make their role difficult to resolve in practices and therefore be left unstated or unaccounted for. This can be seen to be a consequence of moving beyond a human-centred approach. Yet despite these challenges, we still see the need to attend to and account for these material elements.

In our design research investigations of the relations between us and things, we developed a methodology we call material speculations (Wakkary et al. 2015) in which we design artefacts that are crafted to embody research questions or propositions to be lived with over long periods of time. We refer to these artefacts as *counterfactual* because they are designed against current norms of design artefacts as a way to elicit or make visible new phenomena. Inquiries emerge from the lived-with experience and observed existence of the counterfactual artefacts.

Material speculations support our thing-oriented or non-human-centred investigation of the role of things in practices. At the centre of this methodological approach are counterfactual artefacts that are designed specifically for the inquiry. On a general level, the counterfactual artefacts are crafted to be simultaneously familiar and strange. This contradictory nature of the design opens up matters to be investigated by being accepted within a home while equally resisting stable relations to practices. This duality creates actions and elicits reflections on the role of the thing in the home that are productive to study. We discuss two counterfactual artefacts we designed for such inquiries. First, the *Morse Things*, which are sets of ceramic bowls and cups that communicate to each other over the Internet and express these communications sonically through Morse code. As ceramic bowls and cups, they readily become part of the home, yet oddly, the artefacts message each other without human interference. The second counterfactual artefact, the *table-non-table*, is a stack of paper on an aluminium chassis that infrequently moves for short durations. As an object, it is similar in nature to a small table yet it moves on its own. In this context, it's important to state that the presented counterfactual artefacts are purposely designed with weak use goals or links to particular practices (e.g. the *table-non-table* furnishes a home) and are simultaneously designed to problematise these links (e.g. the *Morse Things* communicate in Morse code).

In previous work, we have arrived at various concepts pointing to how interaction with things in addition to direct human use is incremental, of indirect or even no purpose, and revealed by configurations with other things (Wakkary et al. 2016). We have explored ideas in which digital artefacts are designed without concern for human use or for discrete practices, yet these artefacts are adopted into forms of everyday living (Odom and Wakkary 2015). We have described portions of this phenomenon in the *Morse Things* study as a matter of withdrawal, in which all things are not readily transparent to our human understanding and perception, creating a gap between things and us (Wakkary et al. 2017). Later in this chapter, we elaborate on these ideas through our *table-non-table* study and further articulate thing relations as a matter of displacement, referring to the notion that relationships with things can be visible solely through traces and displaced effects of non-human and human actions over time.

Withdrawal and Multistability of the *Morse Things*

As we discussed earlier, understanding experience from the perspective of things is fraught with intelligibility issues that make this a difficult task for people. In response to the alien nature of things, there is a tendency to relate to these non-humans as surrogate humans, that is, to anthropomorphise. Latour (1987) argues that anthropomorphism is not in itself a problem and in fact can productively frame things on a continuum of delegation between human and non-human elements in which artefacts take on human functions. However, complicating this symmetry between humans and non-humans, non-human perspectives can be said to 'withdraw' from human understanding into a non-human world that humans can neither fully comprehend nor articulate (Bogost 2012; Verbeek 2005). In addition, non-human worlds are formed in a configuration of materials and performances rather than language (Baird 2004; Bennett 2009). Withdrawal underlines the difference in materiality and existence between us and things. However, here we consider withdrawal within the confines of hybridity between humans and technology within a postphenomenological subject-object schema (Verbeek 2005) or the mutual constitution of diverse human and non-human entities within practices (Shove et al. 2007). In this case, withdrawal in our view is a condition and an inescapable aspect of the hybrid relation between humans and non-humans. In this sense, we assume that if a thing is intelligible, even in a limited form, it is of a hybrid relation or of a practice, even if the particular practice is not so readily evident.

This characterisation of the relations of things echoes the concept of *multistability* from postphenomenology. Don Ihde (1990, 15) describes how an artefact can hold multiple simultaneous variations depending on the 'active perceptual engagement' of an embodied observer. The multistability of artefacts makes for multiple variations of the same thing to potentially manifest differently dependent on the particular embodied perspectives embedded within given material-social-cultural dimensions. For example, Ihde explains that in the case of an optical illusion like the Necker cube, there is a range of variant interpretations. Similarly, as we will soon discuss, a *Morse Thing*, a ceramic bowl with embedded technology, may switch in an instant from being a bowl for cereal to a communicative device back to a cereal bowl. Considering multistability in relation to withdrawal, a less stable foundation emerges for tracing social practices through things. For example, the question arises which variation or variations are taken up into practice? More fundamentally, when we say we are following the thread of material elements to practices, is it a given variation from the embodied perspective of a human observer or the adoption of the embodied perspective of things in which all its potential variations are inseparable?

The *Morse Things* are sets of ceramic bowls and cups (see Fig. 8.1) that communicate to each other over an Internet connection. Over time, the conversation of the *Morse Things* and their degree of connectedness on the network can evolve in degrees of 'awareness' on the part of a *Morse Thing*, from being alone, to being a pair of things, to being a group of things, to being part of a larger network of things. The *Morse Things* mostly sleep (computationally speaking) and wake at random intervals during the day at least once every eight hours. Upon waking, a *Morse Thing* will send and receive messages to and from other *Morse Things* in

Fig. 8.1 A set of *Morse Things*. ©Everyday Design Studio

its set. The messages sent by each *Morse Thing* are in Morse code and simultaneously expressed sonically and broadcasted on Twitter. The *Morse Things* can be used like any other bowl or cup for eating, drinking, and storing items and so are available to any number of domestic practices. We designed and fabricated six sets of *Morse Things* each including a large ceramic bowl, a medium bowl, and a cup. The form of each *Morse Thing* is made of ceramics that is shaped around the embedded electronics, signalling the design intention to create a fusion of computation and everyday objects.

We deployed the *Morse Things* in six households in Vancouver, Canada, for six weeks. Following the deployment, we conducted a workshop with the study participants to discuss the role of the *Morse Things* and ultimately the gap between things and people.[1] Here, we reflect on the experiences of our participants of living with the *Morse Things* (also see Fig. 8.2) and how these revealed aspects of the withdrawal and multistability of things.

Our study provided details on experiencing the gap between humans and things, revealing the idea of how things recede from and shift within our understanding yet maintain a human relation. It is important that while much of the experience of things is beyond our grasp, this perspective is not entirely invisible. For example, in a presentation during the

[1] See a comprehensive report and discussion of this deployment study in Wakkary et al. (2017).

Fig. 8.2 The *Morse Things* in participants' homes. ©Everyday Design Studio

workshop, participants who were parents of a four-year-old emphasised that their son was the family member who could most easily relate to the *Morse Things* because 'he is already in that space, making Lego and doing things.' This signalled the idea that their son could occupy a different world among things through play and imagination. Such a world is distinct or withdraws from our everyday sensible reality yet is still very much a part of it. Relatedly, two participants separately and consistently throughout our workshop compared the *Morse Things* to pets and teenagers signalling familiar relationships that at times are very unfamiliar and inaccessible to us: 'I think what they do is make us aware that there's other things going on that we have no idea about, like with the teenagers. [...]. I don't know what the cat is doing when I'm sleeping or what my kids are doing.'

In one instance the *Morse Things* were compared to a painting. Like a painting, one does not fully understand a painting yet one forms a special relationship that spurs ongoing reflections and interpretations despite knowing these can never be resolved. Unlike paintings, *Morse Things* express more clearly a relational autonomy that seems to 'interrupt' or emerge into our lives on their own accord shifting our perception of the things. As one of our participants noted: 'You're not really trying to put it [a painting] to any purpose, whereas the bowl … sometimes you are trying to put it to purpose and then it just interrupts you and is like "hey, figure me out".'

Displacement

The discussion of withdrawal and multiplicity of things leaves an unstable ground in its wake from which to consider and observe the role of things in practices. Yet, despite this instability and ambiguity, humans reliably establish commonalities and interactions with things on a constant basis, from the mundane to the critically important. In ToP this allows, with some forceful analytical persuasion, for the integration of things (or their multistable variations) into a practice. As Shove and her colleagues point out, social practice theorists have taken 'endless liberties' in interpreting the role and definition of elements of practice as a trade-off for developing a productive method for understanding practices: 'We have classified cast-iron stoves, skateboards, ink, infrastructures, bicycles, and wing mirrors as material elements when it suits us, paying little or no attention to their different and distinctive qualities' (Shove et al. 2012, 121). Understandably, the boundaries of a material element can be fluid and the heterogeneity among them from scale to materials to types can seem endless. However, for designers attention is paid to the distinct qualities within a given type of material element. For example, there are a myriad number of unique skateboards that have been designed and are yet to be designed. In this sense, we aim to account for the particularities of a given thing within ToP while accepting the incomplete knowing of things. As a starting point to understand practices through the particularities of things, multiplicity, and withdrawal, we propose the idea of *displacement*.

Displacement refers to how the negation or absence of a thing or action can expose unnoticed or undeveloped related phenomena of that thing.

When talking about non-human elements, understanding a thing through everything but that thing helps reveal the relations that are bound up with it. Displacement does not define a thing, its direct use, or the practice it is a part of, because we cannot fully know. Rather, it describes how the thing relates to the world and the configurations of the world it is a part of. It highlights how artefacts exist in our world and how they shape our everyday practices yet remain only indirectly present.

An example of displaced things and their relations with people is a person encountering a stained table with an empty glass of wine upon it. The person can see evidence of relations in the things present (a table, a wine stain, empty glass of wine, and the absent person who drank the wine), but they may have not experienced or witnessed the particular practices that initially brought them together. Evidently, these things were taken up in a practice but the particulars are elusive. Was this wine stain produced during a romantic evening or was her roommate simply enjoying a glass of wine in solitude? Despite what we do not know, we do know that this displaced mediation is evidence of the saying and doings with things.

We elucidate the idea of displacement below through a discussion of the deployment of our *table-non-table*, which is a stack of paper, the size of a small coffee table, supported by a motorised aluminium chassis on wheels (see Fig. 8.3). The chassis and motors are strong enough to support stacking heavy objects on it, including a person sitting or standing on it. On occasion, the *table-non-table* moves for a brief duration of five to twenty seconds. There are close to 1000 sheets of paper per *table-non-table*. Each sheet is held in place by a square die-cut hole that allows it to be stacked around a square aluminium post in the centre of the object but the sheets can easily be removed and used. The almost imperceptible movement is random and stays within an area of less than a square metre.

The *table-non-table* as a material speculation is a counterfactual artefact in that it resembles a small table but counters what might be considered familiar qualities of a table as it moves on its own accord. It embodies the proposition that an unfamiliar or strange digital artefact can fit comfortably in a home in ways we do not explicitly understand, yet foster or make visible the ways a thing may be taken up into domestic practices.

Fig. 8.3 The *table-non-table*. ©Everyday Design Studio

The *table-non-table* was deployed in four households from three weeks to five months. We asked participants who lived with it to post photos and comments about their experiences to a blog and also interviewed them. Our interest was mainly in the reporting of people's subjective encounters with the *table-non-table*.

The Displacement of the *Table-Non-table*

In many of the photos and comments from our deployment studies, the *table-non-table* was only indirectly present but still existed amongst many other actors, portraying indirectly an absorption into practices. In past work, we have described a common type of engagement with things as *intersections*, incidental and even unnoticed encounters with things, and *ensembles*, which are configurations of things (Odom and Wakkary 2015; Wakkary et al. 2016). Intersections accumulate over time embedding the *table-non-table* in an ensemble of artefacts. The more the *table-non-table* seemed to fit within a home, the more it appeared to fade into the background by becoming a part of more complex and diffuse ensembles of things—evident more as a matter of displaced actions or consequences. In Fig. 8.4 the *table-non-table* fits comfortably in this collection

Fig. 8.4 The *table-non-table* among other things in a home. ©Everyday Design Studio

of non-digital things that includes drying laundry, a cast-iron teapot, an objet d'art on the windowsill, and a footrest. How it became part of this ensemble of things is not clear.

Investigations with the *table-non-table* allowed us to see how, through a series of ongoing intersections that result in ensembles of non-human things and humans, things are present in hybrid relations of practice albeit in ways that are displaced.

For example, one participant did not necessarily witness the actual movement of the *table-non-table* but commented on the 'crookedness,' since it would not remain parallel to the couch it was placed alongside: 'my architect eyes were unhappy to see that the thing was always crooked and not parallel to the couch.'

In another situation, the *table-non-table* was an object of curiosity for the household cat who slept on it, clawed it, and played with the loosened paper (see Fig. 8.5) that was later followed by family members making large decorative snowflakes with the paper. Similarly, in another household, participants discovered and marvelled at how their pet rats would 'sneak' paper from the *table-non-table* to build structures inside a play tunnel (see Fig. 8.6) and under their couch which led to the participants 'supplying' *table-non-table* paper to the rats.

In a last example, the paper was separated from the *table-non-table* and stored under a side table and the chassis resting in an open floor space in a living room. Here as well, the particularities of what made up this scene are unclear, and the changed role of the table-non-table is not easily resolved—is it still understood as a table in its deconstructed state, or is it transformed into another thing for another practice?

In these examples, we can see how the *table-non-table* is part of a series of isolated intersections or a string of intersections that threads through

Fig. 8.5 Cat playing with the *table-non-table*. ©Everyday Design Studio

Fig. 8.6 Rats building structures with the *table-non-table* paper. ©Everyday Design Studio

various practices. Here the doings and sayings with things are muted and fragmented but not absent. Displacement can be seen as the material traces of human and non-human actions that configures things, non-humans, and humans in particular if not in conceptually identifiable ways. In this case, our thing, the *table-non-table*, is not viewed as an object on its own but is rather one element bound up with other elements in an ensemble of things integrated into bundles of diverse practices, as Shove and colleagues might argue (Shove et al. 2012), that in our view defies disentanglement.

Conclusion: Things in Practice

Our aim in this chapter was to show how concepts from philosophy of technology and our design research informed our concept of displacement that describes and in part addresses the challenges of tracing things in relation to practices. We discussed how the very ground on which we aim to understand things in practices is less than stable as informed by the philosophical notions of withdrawal and multistability. Withdrawal describes the condition in which non-human worlds, the lives of things, are neither fully perceivable nor comprehensible to human understanding (Bogost 2012; Verbeek 2005). The postphenomenological concept of multistability reveals how artefacts can hold multiple simultaneous variations dependent on the particular embodied perspectives and social-cultural material in which observers of things are embedded (Ihde 1990).

As a result, the visibility of practices is diffused and a matter of residues, traces, or displaced effects of human or non-human actions in what we labelled as displacement. Displacement describes how things take part in the configurations of the world without humans fully understanding that thing, its direct use, or the practice that it is embedded within. In many respects, displacement shifts ideas of practice in order for us to bring things or material elements into greater relief and contrast. As a consequence, practices move to the background of what can be analysed. At best, we can say that practices-as-entities give way to detailed fragments of the performances of practices. Additionally and in return, we have assumed that things that have an ongoing place and effect in everyday life are integrated into practices even if we cannot identify the practice. Within the idea of hybridity, it is difficult for us to consider things outside of practices or hybrid relations. The previously mentioned condensation on a mirror, mould on the bathroom wall, decorative porcelain tiles, batteries from a Bluetooth speaker, or a ceramic ashtray should also be taken seriously when understanding how things co-shape the social every day.

Our efforts to move beyond a human-centred understanding of things in practice moves the decentred subject in ToP (Strengers 2014; Reckwitz 2002) a step further by asserting a hybrid relation between human and non-human entities of practice in which each is distinct. In conclusion, even when things or non-human particularities are not within our grasp and perception, there is still room for them to exist and have effects on our everyday life and the practices of which everyday life is comprised.

References

Baird, Davis. 2004. *Thing Knowledge: A Philosophy of Scientific Instruments*. Berkeley, CA: University of California Press.

Bennett, Jane. 2009. *Vibrant Matter: A Political Ecology of Things*. Durham, NC: Duke University Press.

Bogost, Ian. 2012. *Alien Phenomenology, or What It's Like to Be a Thing*. Minneapolis, MN: University of Minnesota Press.

Desjardins, Audrey, and Ron Wakkary. 2013. Manifestations of Everyday Design: Guiding Goals and Motivations. In *Proceedings of the 9th ACM Conference on Creativity & Cognition (C&C '13)*, 253–262.

Ihde, Don. 1990. *Technology and the Lifeworld: From Garden to Earth*. Bloomington: Indiana University Press.

Latour, Bruno. 1987. Where Are the Missing Masses? The Sociology of a Few Mundane Artifacts. In *The Social Construction of Technological Systems: New Directions in the Sociology and History of Technology*, 225–258. Cambridge, MA: MIT Press.

———. 1993. *We Have Never Been Modern*. Cambridge, MA: Harvard University Press.

Odom, William, and Ron Wakkary. 2015. Intersecting with Unaware Objects. In *Proceedings of the 2015 ACM SIGCHI Conference on Creativity & Cognition (C&C '15)*, 33–42. ACM Press.

Reckwitz, Andreas. 2002. The Status of the 'Material' in Theories of Culture: From 'Social Structure' to 'Artefacts'. *Journal for the Theory of Social Behaviour* 32 (2): 195–217.

Rosenberger, Robert, and Peter-Paul Verbeek. 2015. A Field Guide to Postphenomenology. In *Postphenomenological Investigations: Essays on Human-Technology Relations*, ed. Robert Rosenberger and Peter-Paul Verbeek, 9–41. Lanham: Lexington Books.

Schatzki, Theodore R. 1996. *Social Practices: A Wittgensteinian Approach to Human Activity and the Social.* Cambridge, MA: Cambridge University Press.

Shove, Elizabeth, Matthew Watson, Jack Ingram, and Martin Hand. 2007. *The Design of Everyday Life*. New York: Berg.

Shove, Elizabeth, Matt Watson, and Mika Pantzar. 2012. *The Dynamics of Social Practice: Everyday Life and How It Changes*. Los Angeles, CA: SAGE.

Strengers, Yolande. 2014. Smart Energy in Everyday Life: Are You Designing for Resource Man? *Interactions* 21 (4): 24–31. https://doi.org/10.1145/2621931.

Verbeek, Peter-Paul. 2005. *What Things Do: Philosophical Reflections on Technology, Agency, and Design*. University Park, PA: Penn State Press.

Wakkary, Ron, Audrey Desjardins, and Sabrina Hauser. 2016. Unselfconscious Interaction: A Conceptual Construct. *Interacting with Computers* 28 (4): 501–520.

Wakkary, Ron, Audrey Desjardins, Sabrina Hauser, and Leah Maestri. 2013. A Sustainable Design Fiction: Green Practices. *ACM Transactions on Computer-Human Interaction* 20 (4): 23:1–23:34.

Wakkary, Ron, and Leah Maestri. 2008. Aspects of Everyday Design: Resourcefulness, Adaptation, and Emergence. *International Journal of Human–Computer Interaction* 24 (5): 478–491. https://doi.org/10.1080/10447310802142276.

Wakkary, Ron, William Odom, Sabrina Hauser, Garnet Hertz, and Henry Lin. 2015. Material Speculation: Actual Artifacts for Critical Inquiry. *Aarhus Series on Human Centered Computing* 1 (1): 12.

Wakkary, Ron, Doenja Oogjes, Sabrina Hauser, Henry Lin, Cheng Cao, Leo Ma, and Tijs Duel. 2017. Morse Things: A Design Inquiry into the Gap between Things and Us. In *Proceedings of the 2017 Conference on Designing Interactive Systems (DIS '17)*, 503–514.

9

How Software Matters: Connective Tissue and Self-Driving Cars

Janine Morley

Introduction

In its own right, software has become a topic of social science research, not just as a black-boxed component of the devices, infrastructures, and systems that it enables. By unpacking and peering into the processes by which software is made and by which it operates in everyday life, researchers are exploring the distinctive 'work' that software does in the world: they are asking how "software matters" (Kitchin and Dodge 2011, 3). To date, this has included research into algorithms and their roles in political, personal, and work life (Gillespie et al. 2014; Zarsky 2015; Pasquale 2015; Beer 2017), the production of software (Mackenzie 2006), its introduction and embedding into organisational settings (Pollock and Williams 2008), and its role in the production of urban space (Thrift and French 2002; Kitchin and Perng 2016). Many writers have also grappled with the nature of software itself and the distinctive challenges it poses for

J. Morley (✉)
Department of Sociology, DEMAND Centre (Dynamics of Energy, Mobility and Demand), Lancaster University, Lancaster, UK
e-mail: j.morley@lancaster.ac.uk

C. Maller, Y. Strengers (eds.), *Social Practices and Dynamic Non-Humans*,
https://doi.org/10.1007/978-3-319-92189-1_9

thinking about technological agencies (Thrift 2004; Leonardi 2010; Berry 2011; Kitchin and Dodge 2011). For instance, software is seen to have dynamic, almost living, qualities that render the world "increasingly lively" (Thrift 2004, 461).

For theories of practice, software is a challenging 'material' to conceptualise and study. This is, firstly, because some forms of software 'enliven' and automate processes in complex ways that reduce, remove, or otherwise de-centre the active and practical doings of people. Yet practice theories, such as those of Schatzki (2002, 2010) and Shove et al. (2012) on which I focus in this chapter, emphasise direct bodily activity as the very *basis* of practices, and therefore of social life, stability, and change. So how is software, and the automated work it does in the world, to figure within these approaches?

The second challenge for theories of practice arises in the unusually dynamic nature of software itself: many forms of software can change in significant ways as they are used. In particular, software can be responsive to aggregate, past interactions or live data from beyond the immediate environment, bringing this to bear within situated performances. It can be updated remotely, globally, and quickly in ways that significantly change what objects or systems do. Some forms of software also 'learn' by themselves, as code is re-written through machine learning, without immediate oversight from people. In many applications of practice theory, particularly within studies of consumption, there is a tendency to define 'materiality' as comprised of discrete, relatively static and bounded objects: often limited to the "things handled in a social practice" (Reckwitz 2002, 221). Yet software is not well-bounded, nor is it tangible or fixed in the way that objects like cups, bicycles, or buildings are.

In addressing these two related challenges, this chapter argues that Schatzki's (2002, 2010) and Shove et al.'s (2012) versions of practice theory are not necessarily at odds with the idea of a dynamic materiality that is 'active' beyond moments of direct human practical activity. Indeed, such material doings, happenings, or processes often provide necessary conditions for performances. Moreover, the apparent degree of autonomous 'performance', or the work that software does, appears less independent from practices once it is considered as part of wider complexes (Shove et al. 2012) or systems of practice (Watson 2012). If the aim is a fuller consideration of the work that software does, it is therefore necessary to explore how software is positioned in relation to a range of practices within such complexes. This shifts the focus beyond the single practices which appear to be 'taken over' and 'carried on' by software technologies and thereby the question of whether such technologies are performing these practices.

To develop an alternative, and relational, response to the automating and dynamically adaptive qualities of software, the chapter explores how software acts as a kind of connective tissue between practices: implicated in how practice complexes reproduce and extend over time. In this light, software appears to play a distinctive role compared to more fixed, object-like technologies, raising new lines of enquiry into how software matters.

Overall, the chapter responds to two key themes of the present volume: the question of how to account for 'non-human performance' within theories of practice and the relations that dynamic 'materials' facilitate between practices. The first section of the chapter delineates different types of software and identifies a handful of qualities that are of particular interest when Internet-connected software is integrated into familiar everyday objects. These include (a) the data-based extension and re-making of networks which link domestic and private practices to those of service providers, regulators, and other organisations; (b) the responsive adaptivity of software; and (c) the fluid and rapid ways in which it circulates. This is illustrated through the case of self-driving and connected vehicles. The chapter then moves to the question of whether, and how, such qualities can be accommodated within theories of practice. After arguing that this is possible by attending to the relations between software and practices, the third section explores one approach to this: how software might be conceptualised as part of the 'connective tissue' through which practices hang together and extend over time. Finally, the chapter concludes with reflections on how this may help to orient empirical investigations of how software comes to matter, including understandings of large-scale, systemic change.

Soft Dynamics and Self-Driving Cars

Software infuses everyday life in diverse ways and through many different kinds of devices, infrastructures, and assemblages that "mediate, supplement, augment, monitor, regulate, and ultimately produce collective life" (Kitchin and Dodge 2011, 9). To help situate and differentiate these varied roles, Dodge and Kitchin (2009) propose a taxonomy of 'coded' objects in which software is embedded. This includes (a) objects such as cookers that do not depend on code for their operation but are 'peripherally coded' enabling supplementary functions like timers and (b) objects, referred to as 'codejects', that only function through code in more or less extensive and connected ways. Some codejects, such as USB memory sticks, have firmware that is hard-wired and static; some codejects, like DVD players, are programmable but in relatively fixed ways; and other codejects, like air conditioning units, detect and respond to their immediate environment. Further coded objects, referred to as 'logjects', monitor and record their own use. Some of these, like MP3 players, are relatively stand-alone, whilst others, such as smartphones, are more dependent on networked connections in order to function fully. This latter category are 'networked logjects', and it is through them, in particular, that software transgresses the boundaries of otherwise self-contained objects, connecting them and traces of their use to networks that extend, often instantaneously, over space.

With the arrival of the Internet of Things (IoT), code is increasingly folded into existing categories of everyday objects, like thermostats, TVs, and cars (Macrorie et al. 2016), endowing them with a range of properties (and concerns) previously restricted to computer-like devices, such as smartphones and laptops, and the online services to which they allow access. So although software already matters in many ways, the extension of software into further realms of everyday life, and into previously well-bounded objects, raises new questions as to how practice theoretical approaches might conceptualise and investigate software. Quite unlike the object-like artefacts that applications of practice theory are more used to dealing with, these new kinds of software are associated with a range of 'softer', more fluid and responsive dynamics.

In this chapter, I illustrate these dynamics by focusing on a particular networked logject: the car. In itself, the integration of software into cars is nothing new: most cars are already codejects which depend on software to 'drive' functions like power-assisted breaking, diagnostic functions, and the navigation, entertainment, and information 'centres'. But as cars are increasingly designed to connect directly to mobile communications and Wi-Fi networks, this allows not only for Internet access to be provided for devices taken into the car but also for diagnostic, navigation, and entertainment systems to collect, send, and receive data via the Internet.

In some cases, connected tracking devices are added to cars and other vehicles for insurance, safety, navigation, and logistics purposes. For instance, in the European Union, new cars sold from 2018 onwards must be fitted with an emergency tracking system, called eCall, that will detect when an accident has occurred and automatically call the emergency services, sending data about the vehicle's location, direction of travel, and time of the crash (European Commission 2015). It is likely that this data will then be subject to a range of algorithmic operations, for instance, to predict the severity of injuries (Wallace 2017).

Car companies are also starting to collect data from the vehicles they sell; but exactly what data can be unclear. Tesla, one of the most prominent companies in the development of self-driving cars, releases selected data in the event of accidents to show that the car itself was not at fault (Thielman 2017). By implication, such data as is ordinarily captured and sent 'back' to Tesla provides a sufficiently rich picture of how each of 'their' cars is being driven and in what road conditions as to diagnose blame. In Europe, where data protection laws are more stringent, Audi is also starting to collect data from cars in the company's own fleet, before voluntarily enrolling wider populations.

But why are car manufacturers like Tesla and Audi collecting data from connected cars? It seems it is essential to the development of 'intelligent' vehicles. For instance, Audi plans to develop 'personal assistants' that will make a range of suggestions, depending on traffic and weather conditions about when to leave for work or to call home if you're likely to be late (Gitlin 2016). Machine learning is key to these aspirations: it will allow software in the vehicle to monitor and infer a pattern of use as a basis for generating personalised suggestions for drivers. Just as with 'learning thermostats' like the Nest, this 'tailored support' will also adapt to incoming data about local conditions. The collection of data from current driving performances helps develop the algorithms that could eventually perform such assistance roles.

More hotly anticipated and debated, however, is the development of autonomous or self-driving vehicles. Trials are currently underway in many countries, and connected cars and the data they are already collecting are key to these developments. Tesla's Model S and Model X cars already run 'Autopilot' software which allows for driver-supervised, hands-free driving on motorways. In early 2017, the hardware of the latest Model S was declared to be ready for fully autonomous driving; it is just the software that remains to catch up, as "it is still learning how to drive" (Stilgoe 2017, 1). This is through a process involving deep, machine learning operating on vast datasets generated from many actual performances of driving available to Tesla through their connected fleet. From this data, software can learn the 'rules' and generate algorithms that reliably replicate the same results as real drivers: each driver thereby becomes an "expert trainer for how the autopilot should work" (Fehrenbacher 2015, para. 7, quoted by Stilgoe 2017). This doesn't happen in an isolated car-by-car fashion: rather, according to Elon Musk, the head of the company, "[t]he whole Tesla fleet operates as a network. When one car learns something, they all learn it" (Fehrenbacher 2015, para. 7).

In other words, this is a two-way 'feedback' process. Not only does analysis of vast aggregated data from past driving performances inform the ongoing development of driving algorithms, but those algorithms are then fed back into *all* the cars on an ongoing basis, changing how they function. This responsive adaptation in functionality on a mass scale is possible because software is updateable. Results can be dramatic. Autopilot software was first rolled out in 2015 to tens of thousands of vehicles already in use in the United States, via car owners' Wi-Fi networks: turning them overnight into (partially) self-driving cars (McHugh 2015). When it is ready, the software for fully autonomous driving will be similarly distributed. This mass adaptation is not just a question of the updatability of software per se but also the way in which updates can be circulated. Thanks to near-constant (or routine) broadband connectivity, much software is now designed to be updated automatically and frequently at a global scale (de Decker 2017).

Compared to other non-coded material artefacts, software brings a remarkable dynamism to networked logjects. It means that objects like cars, and how they work, can be re-designed almost on the fly, much like apps or operating systems. Whilst this has become a familiar quality for computing devices, it also brings potential for an increasing rate of change in the 'physical' functionality of everyday objects and how they are used. Software can circulate and change in ways that are quite unlike physical products that have to be designed, tested, manufactured, shipped, stored, purchased, and delivered: in other words, thanks to software, the scale and speed by which networked logjects may change transcends that of more 'concrete', self-contained technologies. Thus, when it comes to developing accounts of change over time, this makes software an exceptionally interesting yet challenging 'material' to work with. It suggests, on the one hand, that software-based innovations like self-driving cars could bring widespread and rapid change to the practices in which networked logjects are embedded: to driving as well as other practices associated with car travel (McDonald 2016; Lin 2016). On the other hand, many changes in software proceed without discernible impact for the practices in which they are embedded; for instance, security requires a process of near-constant change to maintain stability within associated practices.

In sum, this section has highlighted several of the qualities of software that underpin its status as a distinctively dynamic 'material'. First, it highlighted the extended networks in which data and updates may flow to and from networked logjects like connected cars, serving as resources or inputs to practices of driving and a range of related practices, such as emergency services and car design and production. Second, it explained how software can adapt (itself) both locally and across whole fleets of vehicles in response to data about past performances of driving. Thirdly, it highlighted the significance of the rapid and global way in which software can circulate, and with it bring adapted forms of functionality to everyday objects, which may in turn enable rapid changes in practice. I contend that these properties represent modes of feedback that actively contribute to the reproduction of practice complexes in ways that would not be possible without software. Before developing this argument, however, I will briefly turn to the question, pertinent to the themes of this book, of whether and how practice theories can conceptualise the dynamic and apparently 'performative' qualities of software outlined above.

Accounting for the 'Dynamism' and 'Agency' of Software

In what follows, I will argue that the adaptive and fluid qualities of software are not incompatible with the way that practice theorists (in particular Schatzki (2010)) conceptualise materiality. The idea that 'materials' *only* feature as direct, object, or even tool-like elements of practice performances is not necessarily the case. Instead, I suggest that materiality plays important roles as part of complexes of practice, even when not directly embedded in any single practice. This implies that the work that software does in the world—or what it 'performs'—can be understood through its relationality to multiple practices. Argued in this way, the question of whether automated technologies, like cars, perform practices, like driving, does not logically arise. Indeed, in the conclusion, I will suggest that it is something of a distraction from developing relational understandings of the work that software does, that is, its apparent 'agency'.

First, let us consider the status of software as 'a material'. This term is drawn from Shove et al. (2012) and is more properly described as 'a material element'. It refers to a category of component brought together in the performance of practices and necessary to them, alongside skill and meaning. This framework draws on Reckwitz (2002, 208) who notes that "'artefacts' or 'things'… necessarily participate in social practices just as human beings do". In emphasising a practical engagement with the material world over an exclusively interpretive one, Reckwitz's (2002, 212, 210) account of materiality within practice theory highlights 'things' or 'objects' that are 'handled' in practices. Given this, and given a lingering predilection in western thinking to conceptualise materiality in terms of discrete, fixed objects (Ingold 2010), it is not surprising that many who have worked with practice theories empirically have also tended to limit consideration of materiality to objects or things that are used or operated upon as part of a practice.

Yet software stands in stark contrast to this image of a thing or object. Firstly, the ways that it can respond flexibly to predefined data and 'learn' patterns of activity, including how to recreate human-like driving performances, resemble the characteristics of living organisms more than it does of objects (Thrift and French 2002; Thrift 2004). Secondly, software does not exist physically in the way that an object does. It is not physically tangible, yet it does have a physical substrate: it is stored and operates on disks or circuits of some kind. But these physical properties neither define it, nor do they shape the contours of possible actions that can be performed with or by it, which is the case for physical objects. Rather, software "exists in the way that speech or music exist … software when merely written as lines of code loses its essential essence—it's executability" (Kitchin and Dodge 2011, 24). In other words, the essence of software is as a process, a doing: it "does *something* to *something* … it performs functions and processing" (Berry 2008, para. 1).

However, whilst the adaptivity and process-like nature of software are at odds with the image of an object, neither of these characteristics is actually incompatible with understandings of materiality within practice theories. Indeed, Schatzki (2010) describes materiality as arrangements composed of physical, biological, or natural 'somethings' that include "objects, things, properties, *events*, or *processes*" (133, emphasis added). In this account, one way in which materiality forms part of society arises from the biological or physical flows that pass through arrangements and the practices with which they are interlinked (practice-arrangement nexuses). Thus, to Schatzki material arrangements are not as fixed or static as the name suggests: they "are in some sense crystallizations of matter-energy flows" (Schatzki 2010, 137). This resonates with Ingold's (2007, 2010) account of materials as immersed in processes of formation and transformation, as opposed to states of matter (see also Shove et al. 2014). In other words, it is not incompatible with practice theories for a material phenomenon like software to be dynamic, changeable, and responsive to situations, or even to *do* something.

With the possibility for a dynamic materiality that goes beyond concepts of objects or artefacts established, the next step in arguing that software can largely be accommodated within the existing conceptual resources of practice theories is to note that material entities or processes need not be direct parts of the performance of practices in order to matter for, or relate to, them. This is to recognise a variety of roles for materiality beyond simple or direct, tool-like 'elements' of practice. Indeed, there are already a number of moves in this direction, for instance, in thinking through how infrastructures relate to practices (Shove et al. 2015) or the different roles that the same material entities can play as relative to practices (Shove 2017). To Schatzki (2010, 2012), material arrangements and practices are interlinked in many different ways: "practices effect, use, give meaning to, and are inseparable from arrangements while … arrangements channel, prefigure, facilitate, and are essential to practices" (2012, 4) through relations of causality, prefiguration, intelligibility, intentionality, and constitution. In this light, software does not need to figure as a direct constitutive part of practices in order to feature in the way that practices and arrangements hang together and thus how social life transpires.

More specifically, I have argued elsewhere (Morley 2017) that automated machines may be conceptualised as part of processes, like manufacturing or heating, that are organised by, and cut across, many practices. Even if they do not perform a central device-like, resource-like, or even infrastructural role (Shove 2017) in any single practice, automated machines can usually be understood to feature as a necessary component in a complex or system of practices. Like automated machines, software connects to multiple practices even those in which it is not directly handled or 'used'. Thus, the automated work that software allows for can also be accounted for, and studied, as part of complexes of practice. In fact, there is broader movement away from the analysis of single practices to that of multiplicities or complexes of practice and how they change (Schatzki 2011; Shove et al. 2012; Blue and Spurling 2017; Blue 2017).

From this perspective, the work that software does, particularly in automating complex technologies, needs to be understood as a relational outcome with reference to a range of practices. The very ability of software to do work in the world, to "carry, fold, frame and redistribute actions in different environments" (Mackenzie 2006, 423), might be seen as an extension of human agency, perhaps even human bodies (Wallenborn

2013). Mackenzie (2006) refers to this as secondary agency: the capacity to act 'autonomously' yet *on behalf of* a range of other actors, such as the companies, groups, and programmers that build and deploy code, and the people in myriad situations who use it.

The challenge of thinking about what software does, or its 'agency', is therefore one of thinking through relationships with and between practices. In echoing a number of other authors (Barad 2003; Ingold 2007), I would suggest that the apparent 'performative agency' of software, such as when a Tesla car takes over the control of driving along a motorway, is not an inherent attribute of the software, or the car, itself: it does not *have* agency that then allows it to act. Rather, much like human agency, this action is realised and made possible only through the wider practice-arrangement nexuses within which such cars are embedded. Whatever this work or action is, it has been established, is understood, and has effect through practices and the material arrangements to which they connect. In other words, if we want to better understand and conceptualise what self-driving cars, and the software that 'drives' them, actually *do*, it is paramount to pay close attention to how software relates to practices and how it mediates relationships between them.

Connective Tissue: Software Actively Connects Practices

This section briefly explores the idea that at least some of the work that software does in relation to practices can be understood as a kind of connective tissue within practice complexes. The concept of connective tissue is used by Shove et al. (2012, 36) as that which "holds complex social arrangements in place, and potentially pulls them apart". It is developed by Blue and Spurling (2017) in order to analyse how complexes of practice in hospitals change. In particular, they suggest that material-spatial qualities of practices, alongside jurisdictional and temporal qualities, form interconnections within practice complexes.

Based on the case of connected and self-driving cars, outlined above, it is certainly possible to appreciate how software contributes to making, and re-making, relationships between practices in each of these ways. Firstly, by including new data flows within the practice of driving, such as live navigational advice, and by sending new flows of data to car makers and emergency services, connected car software extends the material-spatial co-ordinates of driving. As Dodge and Kitchin (2009, 1359) observe for networked logjects in the home:

> homes are being embedded in real-time, two-way networks so that the everyday nature of domestic practices … can be monitored by service providers with the attendant data being used to profile and social sort customers. The consequence of networked logjects is that homes are increasingly being stretched out across space in networks of greater and shorter length … and they are subject to increasing levels of (corporate) surveillance.

With the arrival of connected cars, the same will be increasingly true for driving and its relation to other practices such as those of emergency services, insurers, and the car industry. Secondly, this could have jurisdictional implications. Commentators note that regulating software-based driving is a very different prospect to that of human driving (Stilgoe 2017). Thirdly, software that enables live sharing of data with third parties and which automates driving has the potential to re-make the temporal connections that hold driving and other practices together. At the simplest level, traffic control measures or emergency services may be able to respond to incoming data over different (faster) time scales, whilst other activities such as working might become possible when commuting by car. In other words, the ability of software to 'rewire' relationships between practices is testament to the way that it 'performs' these connections (or may do in the future).

Shove et al. (2012, 17) propose that such connections within and between practices might also be understood as "circuits of reproduction through which one performance relates to the next and … through which practices shape each other". Between practices, this shaping is referred to as 'cross-referencing'. This denotes how multiplicities of practices (bundles and complexes) are held together, and might include temporal co-ordination, narrative-based organisation, or the ways that elements co-ordinate and link practices when they are shared, passed between practices, or competed for. They also refer to forms of 'monitoring', recording, and feedback that have constitutive roles in performances and feed forward from one performance to the next. If, and when, such forms of feedback cross-over with the circuits formed between practices, it is possible to envisage how circuits of reproduction may work across complexes as a whole. Understood like this, connective tissue is not just about how practices cohere but also how complexes reproduce and change.

In the first section of this chapter, I suggested that the dynamic characteristics of software (that situate technologies within extended data-based networks, that adapt to past performances, and that can be circulated near-instantaneously and globally) might be understood as forms of feedback within complexes of practice. In general, information technologies may already be implicated in 'rewiring' the circuits of reproduction. In commenting on the concept of global microstructures developed by Knorr Cetina (2005), Shove et al. (2012, 117) suggest that:

> the emergence of such modes may be symptomatic of a more pervasive transformation in the circuitry of society itself. Whether or not this is so, it is clear that technologies and instruments of feedback, broadly defined, are of direct relevance in configuring circuits of reproduction that are, in turn, of equally direct consequence for the survival and transformation of relations between practices and the elements of which they are composed.

Through its roles in communication technologies, software already matters in this way. The extending role of software in everyday networked logjects prospectively takes this even further.

In the case of connected and self-driving cars, software enables new possibilities for 'feedback' in the relations between practices described above. These software-enabled circuits of feedback go beyond the relationships between car 'users' and those who make, insure, regulate, maintain them, or provide emergency assistance. They also include circuits of feedback within practices, and practice complexes, from past to present or future performances. This can be seen in the local machine learning that Audi's personal assistant software will depend upon, and in the aggregated, mass 'learning' by which Tesla's Autopilot software is evolving. Here, analytics performed on past performances are combined with analysis of incoming and live data, either to produce contextually specific 'advice' or to affect control of a car. These circuits extend both across time, connecting past to present performances, and space, bringing to bear traces of data about the many other performances of driving that generate traffic conditions. In other words, software does indeed appear to simultaneously 'perform' multiple circuits of reproduction across 'automobility' complexes or systems (Urry 2004).

Taking this idea further, we might suggest that it is the very re-making of connective tissues surrounding driving through which automated driving is becoming an increasingly real prospect. On the one hand, Stilgoe (2017) argues that the emphasis on machine learning makes self-driving cars a much closer prospect by removing some of the previously assumed need to overhaul infrastructures to accommodate these new kinds of vehicles. The development of autonomous vehicles through machine learning based on data of actual driving performances reduces the field of necessary change to one of technological substitution (or delegation) between human and non-human driver. Yet, on the other hand, this apparent substitution, limited to the task of driving itself, can still be seen as a process of transformation in the connective tissue of automobility complexes. This is because it depends on making new relations of feedback between past performances of driving and the development of software, and in turn this depends upon new and different relations between drivers and car makers as observed in the case of Tesla and their access to vast sets of data on what drivers do.

In summary, when it comes to accounting for the software embedded in networked logjects, this section has argued that it is helpful to trace the modes of 'feedback'—between practices and between performances—that forge connections across complexes of practice. Indeed, I have suggested that this could be a useful approach for analysing the very development of self-driving technologies. These modes of feedback are an important aspect of the work that software does in the world, and it is worth stressing that this is not passively achieved, but actively 'performed' through the flows of data, software updates, analytics, and algorithmic processing that make software such a dynamic phenomenon.

Conclusion

In drawing on self-driving and connected cars as an example of how software can matter in new ways, my aim has been twofold. First, I have hoped to set out and respond to some of the qualities of software that make it such an intriguing 'material', particularly for theories of practice that have hitherto been relatively limited in their conceptual and empirical treatment of material elements or arrangements as primarily consisting of a set of discrete, well-bounded, tangible, and static objects. I have argued that this understanding of materiality is not inherent or necessary to practice theories. Rather, alternative and more promising directions are already present and emerging: in Schatzki's (2010) description of material arrangements including processes and events, in Shove et al.'s (2015) discussion of how infrastructures relate to practices, in Shove's (2017) argument that materials occupy roles relative to the practices in question, and in Blue and Spurling's (2017) account of materiality as an aspect of the connective tissue by which practice complexes hang together. In adding to these developments, I argue that the work that software does in the world can be defined in relation to complexes (or systems) of practice.

Therefore, if we have an interest in understanding the distinctively dynamic qualities of software, and the work this makes possible—the 'performative' qualities of software—then the challenge is to understand the multiple relationships within practice complexes through which this work is done. Reducing this to a question of whether self-driving vehicles perform social practices without humans risks loosing sight of the richness of relationships in which driving is embedded. In itself, it is a fascinating topic to explore, as in other chapters in this volume (see Chapters 10 and 11). But, I would suggest that it is not the first question to ask of increasingly sophisticated and 'smart' technological developments like the self-driving car. Indeed, to be occupied only with the practice of driving and the redistribution of performative work within the driver-car assemblage (Dant 2004) is to miss the work that software is doing to make this possible. This can only be understood through transformations in relations across wider systems of practice in which driving is embedded. To this end, it is helpful to conceptualise software as part of the connective tissue through which practices hang together. More specifically, software is plausibly involved in circuits of reproduction at the level of complexes.

This remains a limited and speculative account of how software matters; it is also likely to matter through many other contributions to regulating, sustaining, and reproducing practices. But I hope it indicates a direction of travel. Indeed, there is much more to do in thinking through the conceptualisation of connective tissue and in developing empirically founded analyses of how technologies, and not only software, are implicated in it. This marks a move away from a dominant focus on the direct roles that non-human materiality plays within single practices towards consideration of indirect roles in stitching together practice complexes and the ways in which they reproduce, extend, and change over time. This could help in developing practice theoretical accounts of 'large' social phenomena (Schatzki 2011) and systemic societal change.

I have also noted the parallel that other authors have drawn between software and organic life. Even though software is a technology, and is crafted to carry on human intentionality and thus differs markedly from organic life or geo-physical processes, the approach to exploring indirect relationships in complexes or looser systems of practice, as sketched out in this chapter, may be helpful for analysing a range of other 'non-humans' and their contributions to social life, such as the weather, light, health, or heat. Yet I also hope it is clear that the study of software in cases like the self-driving car may serve as fruitful ground on which to further develop and substantiate the concept of connective tissue, through empirical investigation of circuits of feedback and reproduction.

References

Barad, Karen. 2003. Posthumanist Performativity: Toward an Understanding of How Matter Comes to Matter. *Signs: Journal of Women in Culture and Society* 28 (3): 801–831. https://doi.org/10.1086/345321.

Beer, David. 2017. The Social Power of Algorithms. *Information, Communication & Society* 20 (1): 1–13. https://doi.org/10.1080/1369118X.2016.1216147.

Berry, David. 2008. A Contribution Towards A Grammar of Code. *Fibreculture Journal* 1 (13): 1–1.

———. 2011. *The Philosophy of Software: Code and Mediation in the Digital Age*. Basingstoke: Palgrave Macmillan.

Blue, Stanley. 2017. Institutional Rhythms: Combining Practice Theory and Rhythmanalysis to Conceptualise Processes of Institutionalisation. *Time & Society*. https://doi.org/10.1177/0961463X17702165.

Blue, Stanley, and Nicola Spurling. 2017. Qualities of Connective Tissue in Hospital Life: How Complexes of Practice Change. In *The Nexus of Practices: Connections, Constellations, Practitioners*, ed. Allison Hui, Theodore Schatzki, and Elizabeth Shove, 24–37. Abingdon, Oxon: Routledge.

Dant, Tim. 2004. The Driver-car. *Theory, Culture & Society* 21 (4–5): 61–79. https://doi.org/10.1177/0263276404046061.

de Decker, Kris. 2017. Rebooting Energy Demand: Understanding Software Updates. Accessed August 17, 2017. http://www.demand.ac.uk/rebooting-energy-demand-automatic-software-updates/.

Dodge, Martin, and Rob Kitchin. 2009. Software, Objects, and Home Space. *Environment and Planning A* 41 (6): 1344–1365. https://doi.org/10.1068/a4138.

European Commission. 2015. eCall in All New Cars from April 2018. Accessed September 9, 2017. https://ec.europa.eu/digital-single-market/news/ecall-all-new-cars-april-2018.

Fehrenbacher, Katie. 2015. How Tesla is Ushering in the Age of the Learning Car. *Fortune*. Last modified 16 October 2015. Accessed September 4, 2017. http://fortune.com/2015/10/16/how-tesla-autopilot-learns/.

Gillespie, Tarleton, Pablo Boczkowski, and Kirsten Foot. 2014. The Relevance of Algorithms. In *Media Technologies: Essays on Communication, Materiality, and Society*. MIT Press.

Gitlin, Johnathan M. 2016. Audi Wants Its Connected Cars to Improve the Breed. Accessed August 9, 2017. https://arstechnica.co.uk/cars/2016/09/personal-assistants-and-data-analytics-the-future-of-audis-car-ux/.

Ingold, Tim. 2007. Materials Against Materiality. *Archaeological Dialogues* 14 (1): 1–16. https://doi.org/10.1017/S1380203807002127.

———. 2010. Bringing Things Back to Life: Creative Entanglements in a World of Materials. *NCRM Working Paper*. Realities/Morgan Centre, University of Manchester.

Kitchin, Rob, and Martin Dodge. 2011. *Code/Space: Software and Everyday Life*. MIT Press.

Kitchin, Rob, and Sung-Yueh Perng. 2016. *Code and the City*. Routledge.

Knorr Cetina, Karin. 2005. Complex Global Microstructures The New Terrorist Societies. *Theory, Culture & Society* 22 (5): 213–234. https://doi.org/10.1177/0263276405057200.

Leonardi, Paul M. 2010. Digital Materiality? How Artifacts Without Matter, Matter. *First Monday*. https://doi.org/10.5210/fm.v15i6.3036.

Lin, Patrick. 2016. Why Ethics Matters for Autonomous Cars. In *Autonomous Driving*, ed. M. Maurer, J. Gerdes, B. Lenz, and H. Winner. Berlin: Springer.

Mackenzie, Adrian. 2006. *Cutting Code: Software and Sociality*. Vol. 30. Peter Lang.

Macrorie, Rachel, Mette Kragh-Furbo, and Janine Morley. 2016. Daily Life, Digital Technologies and Energy Demand. *Working Paper Collection*.

McDonald, Rob. 2016. Why Driverless Cars May Make Cities Sprawl Even More. Accessed September 4, 2017. https://blog.nature.org/science/2016/04/20/why-driverless-cars-cities-sprawl-urban-suburban-development/.

McHugh, Molly. 2015. Tesla's Cars Now Drive Themselves, Kinda. *Wired*. Accessed April 20, 2017. https://www.wired.com/2015/10/tesla-self-driving-over-air-update-live/.

Morley, Janine. 2017. Technologies Within and Beyond Practices. In *The Nexus of Practices: Connections, Constellations, Practitioners*, ed. Allison Hui, Theodore Schatzki, and Elizabeth Shove, 81–97. Abingdon, Oxon: Routledge.

Pasquale, Frank. 2015. *The Black Box Society: The Secret Algorithms that Control Money and Information*. Harvard University Press.

Pollock, Neil, and Robin Williams. 2008. *Software and Organisations: The Biography of the Enterprise-Wide System or How SAP Conquered the World*. Abingdon, Oxon: Routledge.

Reckwitz, Andreas. 2002. The Status of the 'Material' in Theories of Culture: From 'Social Structure' to 'Artefacts. *Journal for the Theory of Social Behaviour*: 195–217. https://doi.org/10.1111/1468-5914.00183.

Schatzki, Theodore R. 2002. *The Site of the Social: A Philosophical Account of the Constitution of Social Life and Change*. University Park, PA: The Pennsylvania State University Press.

———. 2010. Materiality and Social Life. *Nature and Culture* 5: 123–149. https://doi.org/10.3167/nc.2010.050202.

———. 2011. Where the Action is (On Large Social Phenomena Such as Sociotechnical Regimes). *Working Paper 1*, Sustainable Practices Research Group.

———. 2012. A Primer on Practices. In *Practice-Based Education: Perspectives and Strategies*, ed. Joy Higgs, Ronald Barnett, Stephen Billett, Maggie Hutchings, and Franziska Trede. Rotterdam, The Netherlands: Sense Publishers.

Shove, Elizabeth. 2017. Matters of Practice. In *The Nexus of Practices: Connections, Constellations, Practitioners*, ed. Allison Hui, Theodore Schatzki, and Elizabeth Shove, 155–168. Abingdon, Oxon: Routledge.

Shove, Elizabeth, Mika Pantzar, and Matt Watson. 2012. *The Dynamics of Social Practice: Everyday Life and How It Changes*. London: SAGE Publications Ltd.

Shove, E., G. Walker, and S. Brown. 2014. Material Culture, Room Temperature and the Social Organisation of Thermal Energy. *Journal of Material Culture* 19 (2): 113–124. https://doi.org/10.1177/1359183514525084.

Shove, Elizabeth, Matt Watson, and Nicola Spurling. 2015. Conceptualizing Connections: Energy demand, Infrastructures and Social Practices. *European Journal of Social Theory* 18 (3): 274–287. https://doi.org/10.1177/1368431015579964.

Stilgoe, Jack. 2017. Machine Learning, Social Learning and the Governance of Self-Driving Cars. *SSRN Research Paper*.

Thielman, Sam. 2017. The Customer is Always Wrong: Tesla Lets Out Self-Driving Car Data—When It Suits. Accessed September 4, 2017. https://www.theguardian.com/technology/2017/apr/03/the-customer-is-always-wrong-tesla-lets-out-self-driving-car-data-when-it-suits.

Thrift, Nigel. 2004. Electric Animals: New Models of Everyday Life? *Cultural Studies* 18 (2–3): 461–482. https://doi.org/10.1080/0950238042000201617.

Thrift, Nigel, and Shaun French. 2002. The Automatic Production of Space. *Transactions of the Institute of British Geographers* 27 (3): 309–335. https://doi.org/10.1111/1475-5661.00057.

Urry, John. 2004. The 'System' of Automobility. *Theory, Culture & Society* 21 (4–5): 25–39. https://doi.org/10.1177/0263276404046059.

Wallace, Nick. 2017. EU's eCall Regulation Sacrifices Safety for Privacy. Accessed August 9, 2017. https://www.euractiv.com/section/digital/opinion/eus-ecall-regulation-sacrifices-safety-for-privacy/.

Wallenborn, Grégoire. 2013. Extended Bodies and the Geometry of Practices. In *Sustainable Practices: Social Theory and Climate Change*, ed. Elizabeth Shove and Nicola Spurling, 146–164. London: Routledge.

Watson, Matt. 2012. How Theories of Practice can Inform Transition to a Decarbonised Transport System. *Journal of Transport Geography* 24: 488–496. https://doi.org/10.1016/j.jtrangeo.2012.04.002.

Zarsky, Tal. 2015. The Trouble with Algorithmic Decisions: An Analytic Road Map to Examine Efficiency and Fairness in Automated and Opaque Decision Making. *Science, Technology & Human Values* 41 (1): 118–132. https://doi.org/10.1177/0162243915605575.

10

Automated Artefacts as Co-performers of Social Practices: Washing Machines, Laundering and Design

Lenneke Kuijer

Introduction

Consider the following event that we can imagine to have recently taken place somewhere in a home:

> Click. He closed the latch and opened the valve to let the water flow into his belly. When he sensed that the right level had been reached, he closed it. He then heated the water to the preferred temperature and triggered the motor to start its gentle pattern of turning clockwise, counter clockwise, clockwise, counter clockwise. When the right length of time had passed, he opened the other valve to extract the water, while initiating the second motor pattern of faster spins in both directions. When he was done, he made the sound and released the latch. Click.
>
> I tried to ask him why he had run the program without anything in his drum, but he didn't know.

L. Kuijer (✉)
Department of Industrial Design, Eindhoven University of Technology, Eindhoven, The Netherlands
e-mail: s.c.kuijer@tue.nl

C. Maller, Y. Strengers (eds.), *Social Practices and Dynamic Non-Humans*,
https://doi.org/10.1007/978-3-319-92189-1_10

Automated artefacts, like this washing machine, are a fascinating type of critter. Over the past century or so, they have made their way into our everyday lives. However, sitting somewhat awkwardly between material elements/arrangements and human performers, they have yet to find their place in theories of practice. Originating from my long-standing interest in social practice theory (Reckwitz 2002b; Schatzki 2002, 2010; Shove et al. 2012) and design, this chapter considers one possible way of placing automated artefacts in the practice-theoretic framework, namely, by considering them as co-performers of practices.

In this chapter, I define automated artefacts as the type that, due to access to energy supply, is capable of performing actions in the absence of people, and, due to their being equipped with sensors, actuators and computing power can make decisions about how to proceed in specific, situated circumstances. Such automated artefacts are routinely studied and developed within the human-computer interaction (HCI) research community that I consider myself part of.

My positioning of automated artefacts as co-performers of practices emerged in response to ponderings over the positioning of 'acting' technologies within practice theories in the work of Morley (2016). Morley argues that when machines 'take over' work that was previously carried out by people, they cease to directly 'participate', as elements, within the performance of practices (Morley 2016, 82). In other words, when an automated artefact such as a washing machine acts outside of the direct presence and awareness of people, its work of washing garments is no longer considered part of the performance of laundering.

While the refinement of extended relations between practices and automated 'background processes' enabled by this positioning is valuable as well as insightful, my previous research suggests that it also hides aspects of the role of automated artefacts, and by extension the role of their designers, in practice dynamics. This research involved a historical case study of daily life on council housing estates in the Northern English town of Stocksbridge from 1920 to 1970 (Kuijer and Watson 2017). Interviewing former estate residents about everyday activities, such as space heating and doing the laundry, in the days before central heating and washing machines exemplified that the details of where, when, how, with what and why things were done helped to gain deeper understanding of the workings of space heating and laundering as practices, and their position within webs of related practices. For example, in Kuijer and Watson (2017), Matt Watson and I argue how the precise way in which fuel is added to a heating system matters for the patterns in which homes become heated over time.

These experiences made me wonder whether opportunities for understanding the role of new technologies and their designers in practice dynamics might be missed by dismissing the details of practice performance once work is delegated to an automated artefact. Today, the mixing of garments with water and soap indeed requires little work and attention from people, but an average washing machine is quite busy with it several times per week. What if, rather than positioning automated artefacts outside of performances of practices, the concept of performance was instead extended to include these 'artificial doings' as part of it? In such a view, automation does not result in the 'dissolution of practices' (Morley 2016, 86), but in a redistribution of roles within them.

This chapter explores these questions and argues that although people tend to be viewed as the sole performers of practices in contemporary practice theory, it is nevertheless both possible and potentially insightful to consider people and automated artefacts as co-performers. In the chapter, I first position the idea of artefacts that perform practices in relation to the conceptual frameworks of Schatzki (2002, 2010), Reckwitz (2002a, b) and Shove et al. (2012). I choose these authors specifically because they explicitly discuss the role of materials in practices. Next, I draw in related social theory from science and technology studies (STS) on distributed agency, in particular the work of Collins and Kusch (1998), Pickering (1993) and Rose and Truex (2000), to specify my interpretation of the agency of artefacts. The core of the chapter consists of a thought experiment in which the concept of co-performance is applied to the analysis of laundering as a situated practice and its dynamics over time. I draw on data from the Stocksbridge case study (Kuijer and Watson 2017), other literatures on practices of laundering and my own experiences and informal knowledge of laundering in the Netherlands. In the conclusion, I raise a range of questions that co-performance triggers, and speculate on how the new perspective it presents on the role of design and designers in social change might inform design practice.

Automated Artefacts as Co-performers in Practice Theory

Because automated artefacts tend not to be discussed as a separate category in contemporary conceptual frameworks of practice theory, the focus in this section lies on the role of materials in relation to practice performance more broadly. I zoom in on Schatzki (1996, 2002, 2010), Shove et al. (2012) and Reckwitz (2002a, b) because these theorists, as opposed to Giddens (1984) and Bourdieu (1977), for example, actively engage with 'the material' in their work and, not surprisingly given their engagement with the material world, have been most influential in technology design and research (Kuutti and Bannon 2014). I will take the work of each of these authors in turn and discuss how they understand the role of materials in their theories and what scope they provide for thinking about artefacts as performers of practices. In the final part of the section, I will make the link back from performance to co-performance by explaining why I believe it is conceptually inconsistent to consider artefacts as performers of social practices in their own right.

Assumed in the following discussion is the distinction between practice-as-performance and practice-as-entity, as well as the recursive relation between them (Warde 2005; Ingram et al. 2007). In summary, what is important for this account is the notion that practices-as-entity are both maintained and transformed through their collective individual performances (Shove et al. 2012).

Nonhumans Display Agency of the 'Performance Type'

For Schatzki, performances entail the doings and sayings of everyday life. People, in his view, are always engaged in this or that practice, or in other words, (practically) all human action is the performance of one or more practices (Schatzki 1996). Schatzki is critical of accounts that 'contend that practices comprise the actions of various entities and not those of people alone' (Schatzki 2002, 71). In his discussion of 'Materiality and Social Life', Schatzki (2010) consistently speaks of the social as comprised of nexuses of material arrangements and *human* practices. However, when he elaborates on agency, equating it to 'doing', he acknowledges it is not restricted to 'bodily human doings', but that 'nonhumans do things as much as humans do' (Schatzki 2002, 191). He then moves on to describe two types of agency/doings: a causal type and a performance type. Agency of the causal type is to make something happen, which can be done by humans and nonhumans alike. The performance type is about carrying out bodily doings and sayings that, in the circumstances at hand, amount to performing the action while at the same time 'carrying on the practice of which it is part' (Schatzki 2002, 191). By adding that 'the extent to which nonhumans can be accorded this sort of agency is an open question' (Schatzki 2002, 191), the floor is open to exploring what nonhuman performance might mean.

Automated Artefacts as Integrators of Elements

For Shove et al. (2012), performing practices involves the integration of elements, generally grouped as 'materials, competences and meanings' (Shove et al. 2012, 14) into unique configurations. The categorisation isn't specific about the roles of people or things in the performance of practices, as human bodies can be viewed as materials, things can have competencies, and meanings can be embodied in both things and humans. While these authors discuss the work of integrating elements solely in terms of human performances—for example, they explicitly refer to 'the integrator' as a human actor (Shove et al. 2012, 22)—they do seem open to the possibility that things might act by mentioning that 'agencies and competencies are distributed between things and people' (Shove et al. 2012, 10).

When, for the purpose of this exercise, conceptualising performance simply as the unique integration of elements in a situation, then automated artefacts can be considered capable of performing practices because of their capability to perform actions outside of the presence and awareness of people. For example, when washing clothes, a washing machine integrates elements of water, soap, garments, electric power, chemical processes, temperatures, movements, durations and sequences into a performance of machine washing that produces cleaned clothes. If taken as an integration of elements, then every performance of a washing machine is unique. The composition of the load of items washed is different each time, as is its level of dirtiness. Maybe the composition and temperature of the water differs slightly, as well as the added amount of detergent. There are clear limits to this narrow framing of machine performance, which I will discuss more towards the end of this section, but for now I make the point that automated artefacts can, to some extent, be viewed as capable of integrating elements into unique performances.

Artificial Body/Minds

Reckwitz summarises the role of people in practice theory as 'body/minds who "carry" and "carry out" social practices' (Reckwitz 2002b, 256). In his discussion of 'the status of the material' (Reckwitz 2002a), he draws on Latour's (1993) idea of hybrids and presents a more 'active' conceptualisation of the role of 'the material' in practices than Schatzki (2010) or Shove et al. (2012). However, in his writings, Reckwitz too ascribes the 'carrying' and 'carrying out' of practices to people alone. His interpretation of performance is human centred in the sense that in all his examples, things are handled *by people* and act as resources *for people* to perform practices. By specifying performance as performing with things, it becomes clear that while he acknowledges things as 'irreplaceable, constitutive elements' that 'enable and constrain the specificity of a practice' (Reckwitz 2002a, 212), it is people that do the performing. However, the possibility of artefacts performing is not explicitly closed down. If for the sake of this exercise we broaden our notion of what a 'mind' is, then the idea of body/minds described by Reckwitz (2002b) could be extended to the hardware/software making up automated artefacts such as washing machines as 'artificial body/minds' capable of routinised performances.

Co-performance

With a bit of stretching, I've briefly drawn out the concepts of nonhuman doings as performance, automated artefacts as integrators of elements, and artificial body/minds from a section of the practice theory literature.

Before moving on to elaborate on the opportunities and implications of considering automated artefacts as performers of practices, I want to make an important addition to my stance. In line with literatures on distributed agency in science and technology studies (STS), I consider the agency of automated artefacts different from human agency. Core to this difference is the idea of intentionality or purpose. Pickering (1993), for example, argues that machines in themselves do not pursue ends; their actions 'are always situated within a space of human purposes, goals, plans' (Pickering 1993, 577). Jones, cited in Rose and Truex (2000, 374) agrees and argues that 'inanimate objects should be allowed agency, but without the particularly human component of intentionality', and Collins and Kusch (1998) argue that because they lack intentions, machines can only 'behave', where people 'act'. In other words, machine washing is only meaningful because people (intentionally) pursue such ends as clean garments, tidy households, 'convenience', 'easy lives' and representative appearances.

Therefore, while doings—and sayings, in the form of sounds and information on displays—can be performed by artefacts as they integrate elements into unique performances, the ends involved remain those pursued by people. Because in theories of practice these ends are considered as part of the practice rather than somehow internal to individual people, it is consistent to say that practices in this sense are indeed always performed by humans. In combination with the proposition developed above, that artefacts can be considered capable of performing practices; this means that automated artefacts, as artificial body/minds, can only be considered as *co*-performers of practices alongside people, rather than as performers of practices in their own right (as explored in Strengers' contribution to this collection). I take the position that without the link to human purposes, artificial performance would be meaningless and therefore lack an essential element to be a social practice.

Understanding Laundering Through Co-performance

Having established a number of concepts that could position artefacts as co-performers of practices while staying within or at least close to the practice-theoretic framework, this section applies the concept of co-performance to find out how it might enhance our understanding of the role of automated artefacts in social change. It will do so by using the concept of co-performance to experimentally analyse laundering, an early and widely studied example of automation in daily life. These experiments are not elaborate or in-depth. Their function is merely to serve as practical examples to illustrate the potential usefulness of co-performance as a concept for understanding relations between everyday practices, technology design and social change.

The section has two parts. The first part zooms in on laundering as a practice that is co-performed by humans and automatic washing machines. The second part zooms out of situated performance towards the dynamics of laundering as a practice with a career of its own and explores how the concept of co-performance sheds a different light on automation as a process unfolding over time.

Analysing Laundering as Situated Co-performance

Performing laundering as a situated practice can crudely be depicted as a sequence of doings, such as in the list in Table 10.1.[1] By specifying which of the doings in the practice are performed by humans and which by automated artefacts, a simple overview is created of a division of roles.

Analysing what is done and how is core to understanding situated performances of social practices, but forms only part of the story, because it does not reveal much about *why* performers act as they do. When taking the details of 'artificial performance' seriously, the question arises of why artefacts do what they do in the way they do it. While most people can probably say something about how washing machines work, inquiry into the details of their workings and the reasoning behind this eventually implicates their designers. As Shove points out in an article on 'technology and practice'—in which she also uses laundering as an example—developers of washing machines have a clear influence on the practice

[1] Such a list of the steps involved in performing laundering can never be complete and varies depending on the situation, this is merely an example composed on a level of abstraction appropriate for supporting the argument in this chapter.

Table 10.1 The co-performance of laundering by a human and an automatic washing machine

Performing laundering	Human	Artefact
Use garment	X	
Put garment in laundry	X	
Compose a load for laundering	X	
Place load in the machine	X	
Add detergent	X	
Select programme	X	
Switch on machine	X	
Heat water		X
Mix load, water and detergent		X
Add rinsing water		X
Rinse out dirt and detergent		X
Switch to spin dryer mode		X
Remove excess water		X
Expose garments to (warm) air	X	
Prepare garments for storage	X	
Store garments for next use	X	

because they 'determine just what it is that the washing machine can do' and therefore have an important say in 'what cleaning means' (Shove 2004, 1061). While Shove stops here, I take the idea that developers design part of the performance of the practice further by speculating on what such an inquiry might involve and render.

When viewing laundering as co-performed, Table 10.1 includes doings performed both by humans and washing machines. This means that, like humans, artefacts embody practical knowledge and know-how. The transcript below of a 1960 English television advertisement for a then new washing machine helps to illustrate what might be highlighted by this perspective:

> *New English Electric Liberator: the dial-a-fabric washing machine that outdates all others. All you do is dial 1 (for heavy wash), 2 (for cottons), 3 (for nylon and terylene), or any of the six different fabric groups. The Liberator does the rest. It heats the water to the right temperature, washes for the right length of time, with a gentle tumble action, which means your clothes will last longer, rinses again and again, spin dries, all in the same tub, and switches off, all this entirely by itself. You can be miles away [footage shows woman and girl at*

grocery store]. And when you come back, your washing has been done for you. That's what dial-a-fabric washing means. (Huntley Film Archives,[2] emphasis added)

The advert reflects the idea of a division of tasks, where there are parts 'you' do and parts the washing machine does. It also reflects the point made earlier that in the case of automation, ends remain human ends as in the expression that what the washing machine does, it does *for you.* Importantly, the transcript highlights that the machine embodies judgments of appropriate performance. In deciding the details of what the machine does and how—heating water to the 'right' temperature, washing for the 'right' length of time—developers of the machine have materialised practical knowledge, that is, how water, detergent and movement work together to remove dirt from garments, into the machine's performance. Because the machine takes part in the performance of laundering, these judgments are subsequently repeated and therefore reproduced in myriad situations, thereby becoming part of the practice-as-entity. Analysing practices as co-performed thus implies specific inquiry into the practical knowledge embodied into the artefact. It also points to the fact that certain qualitative judgments, what is a *right* temperature and length of time, were made outside of situated performance. I will return to this point in the next section, but want to highlight here that an important difference between a co-performance perspective on the role of washing machines and their developers in shaping practices of laundering and the perspective taken by Shove lies in the word 'can' used in the quote above (Shove 2004). Describing the role of the machine's developers as determining what the washing machine *can* do implies that the user, as operator of the machine, has control over what the washing machine *actually* does. Co-performance highlights that although washing machines do nothing without being switched on, once running, they automatically decide how to proceed.

Conceptualising the role of washing machines as co-performers implicates the practices of machine developers in the analysis of situated per-

[2]Transcribed from 'Two Adverts for Washing Machines Film 4764', www.huntleyarchives.com. https://www.youtube.com/watch?v=1PHEdJrqkuM, accessed August 12, 2017.

formances of laundering. In addition, the concept of co-performance points to another type of explanation for the way in which laundering is organised. This involves the analysis of *differences between human and artificial body/minds*. While applying the concept of co-performance to practice analysis, I notice that although it argues for a positioning of automated artefacts as more similar to people than they currently are in theories of practice, what this move actually highlights are *differences* between humans and automated artefacts. Why washing machines do what they do, and why the work of laundering is divided in the way it is, can in part be explained in terms of differences in capabilities between human body/minds and artificial body/minds. For example, judging which garments are dirty and which are best combined into a load is not something (contemporary) washing machines are capable of, while spinning a drum at 1400 rounds per minute is not something humans can do. This idea of different body/minds will be elaborated on in the next section.

Understanding Changes in Laundering Over Time

Important in theories of practice, in particular in the work of Shove et al. (2012), is the idea that practices change over time. This section explores whether and how co-performance might shed new light on how laundering has changed over time and the role automated artefacts have played in this.

Table 10.2 depicts a simplification of how the practice of laundering has changed over the past century, both in terms of changing performances and changing divisions of work between humans and washing machines. The table represents a way of visualising how, over the career of laundering, washing machines have taken over more and more work from humans.[3] This process of cumulative delegation[4] of work to automated artefacts is what I refer to as automation.

[3] At least when considering how single performances have changed. Schwartz Cowan (1983) argues how automation in laundering has resulted in an overall increase in 'work for mother', which happened because of increases in the amounts of laundry and therefore the frequency of performing it.

[4] A term famously applied by Latour (1992) to refer to the transfer of work between humans and nonhumans.

Table 10.2 An overview of automation in laundering with a focus on clothes washing (H = human, A = artefact)

Hand wash	H	A	Twin-tub washing machine	H	A	Automatic washing machine	H	A
Use garment	X		Use garment	X		Use garment	X	
Put garment in laundry	X		Put garment in laundry	X		Put garment in laundry	X	
Compose a load for laundering	X		Compose load for laundering	X		Compose a load for laundering	X	
Heat water	X		Place load in machine	X		Place load in the machine	X	
Mix load, water and detergent	X		Add hot water and detergent	X		Add detergent	X	
Rinse out dirt and detergent	X		Switch on machine	X		Select programme	X	
Remove excess water	X		Mix load, water and detergent		X	Switch on machine	X	
Expose garments to (warm) air	X		Add rinsing water	X		Heat water		X
Prepare garments for storage	X		Rinse out dirt and detergent		X	Mix load, water and detergent		X
Store garments for next use	X		Move load to spin dryer	X		Add rinsing water		X
			Switch on spin dryer	X		Rinse out dirt and detergent		X
			Remove excess water		X	Switch to spin dryer mode		X
			Expose garments to (warm) air	X		Remove excess water		X
			Prepare garments for storage	X		Expose garments to (warm) air	X	
			Store garments for next use	X		Prepare garments for storage	X	
						Store garments for next use	X	

Building on the idea of differences between human and artificial body/minds introduced in the previous section, a focus on co-performance shows how the capabilities of washing machines to perform tasks in the doing of laundering have extended over time. When for a moment generalising from this example, such a view on how practices change highlights the particular capability of artificial body/minds to embody practice-specific, accumulated learning as *a category of artefacts.* While early washing machines were able to simply mix garments with soap and water, their successors are now capable of washing and drying a load automatically. It also highlights a particular difference between human and artificial performers. Human body/minds can embody accumulated learning, and humans have clearly learned new laundering skills over time, such as selecting the appropriate programme and judging which garments are suitable for machine washing. However, as living and biologically reproducing body/minds, they are not as capable of incorporating accumulated learning as artefacts, which can (radically) change shape over different 'generations'.

Holding on to this idea, the following example illustrates how this particular capability of machines, together with the idea that the details of artificial performance matter for how the practice is configured, helps to specify relations between automation and social change.

A 1955 TV commercial for the Easy Washing Machine contains the now outdated slogan:

'Anything that you can wash, Easy does it better'[5]

This slogan uses hand washing as a reference for machine washing. What can be argued from a co-performance perspective is that while this washing machine is claimed (by its developers) to wash better, it actually washes *differently*. It does so simply because its body/mind is different from human body/minds and has different capabilities, strengths and weaknesses. This means that in the process of delegation, a previously human task needs to be *transformed* in order to be suitable for performance by an artificial body/mind. In this process, something is lost and something is gained. For example, while washing machines might be capable of washing garments 'again and again' without complaining, they are not capable of treating each garment slightly differently in response to its specific fabric, stains, shape and social and practical significance in the household.

[5]Transcribed from '1950s TV Commercial for Easy Washing Machine (1955)', https://www.youtube.com/watch?v=6DFaFhjBs9c&spfreload=10, accessed August 12, 2017.

The advert for the New English Electric Liberator emphasises that compared to earlier machines, the Liberator is gentler so that 'your clothes will last longer'. The fact that this needs emphasising illustrates that washing machines, as compared to hand washing, were experienced to reduce the lifespan of clothing. Judging from the fact that even to date the most delicate fabrics and garments are recommended to be hand washed, it can be concluded that garments wear out quicker in the machine.

That automation, while generally viewed as progress, also has downsides is not a novel observation. However, the specification of automation as a process of redefining appropriate performance to fit the capabilities of artificial body/minds seems to present a novel perspective on the role of new technology development in social change. When viewing automation as a process of cumulatively delegating work from humans to automated artefacts, co-performance highlights that this process is fuelled by the unique capability of artefacts to embody accumulated learning over generations. Moreover, it specifies how standards and expectations in these practices are shifting towards dependence on artificial performance as practices reconfigure to favour artificial capabilities.

Building on this idea of automation as involving a process of transforming actions, I would like to close this section with a discussion of a particular form of delegation. Collins and Kusch (1998), in their book *The Shape of Actions: What Humans and Machines Can Do*, work to establish a distinction between *polimorphic* and *mimeomorphic* actions.[6] Polimorphic actions, such as throwing a dinner party, rely on social awareness, while mimeomorphic actions, such as rafting down a river, do not. I now discuss the case of delegating a polimorphic action from a human to an artificial performer.

Other than mimeomorphic ones, polimorphic actions 'involve varying behaviour to carry out the same action in relation to a situation … the appropriate behavioural shape of [which] has to be determined by reference to the society' (Collins and Kusch 1998, 33). Polimorphic

[6] Collins and Kush do not distinguish practices as a unit in the same way that the practice theorists drawn on in this chapter do. Instead, they talk about actions. In some cases these actions could be viewed as practices, but mostly they discuss 'smaller' units that form only part of a practice performance. Acknowledging that the distinction between practices and actions is in part a matter of framing, I clarify that in this chapter I focus on the practice of laundering and view any doings performed to accomplish it as actions.

actions require social awareness. Collins and Kusch (1998) argue that humans build up social awareness over their lifetime through the process of *enculturation*, a concept developed in Anthropology. More engagement with this body of literature is required to make a substantiated point here, but based on various definitions of enculturation that describe it as a set of sophisticated skills of the situated interpretation of tacit, implicit and ambiguous information (Kottak 2005; Poole 1994; Williams 1975), I would argue that automated artefacts cannot be considered enculturated. It is questionable, moreover, whether artefacts can ever become enculturated because they lack the kind of receptiveness to be raised into competent participants of a society that infants are born with. This would mean that they are fundamentally and chronically ill equipped to perform polimorphic actions.[7]

Having said this, it is nonetheless clear, from the example of the washing machine as well as others described in Collins and Kusch (1998), that polimorphic actions can be delegated to automated artefacts. In laundering, for example, the mixing of a load of laundry with water and detergent used to be a polimorphic action, because the treatment of individual garments within a load according to their social significance—for example, treating a favourite sweater with more care than an old apron—requires social awareness. However, with the delegation of the task of mixing garments, water and detergent to automatic washing machines, it became a mimeomorphic one where all garments in a load are treated in the same way.

From a co-performance perspective, the delegation of polimorphic actions to automated artefacts means that socially complex judgments about appropriate ways of proceeding need to be materialised into the machine performance during the design process. This means that these are transferred from the situated performance to practices of development. Materialising such judgments is a challenging task, because, as Collins and Kusch argue, it is 'not possible to provide a recipe for doing [polimorphic actions] the right way that could be followed by someone [or something] who did not already understand the society in which they are embedded' (Collins and Kusch 1998, 33). When considering artefacts

[7] There are also mimeomorphic actions that machines are not capable of, such as, for example, rafting a river, but these are different in the sense that in principle, artificial body/minds could be developed to perform them.

as incapable of enculturation, this implies that even when the designers of an automated artefact are enculturated, it remains impossible for them to equip an artificial body/mind with the recipe for appropriately performing polimorphic actions.

So when actions that rely on social awareness are delegated to artefacts, artificial performances may become inappropriate, or even unacceptable in the diversity of situations in which such artefacts end up. The toddler switching on the washing machine may have her own reasons to do so, but to perform a full washing cycle without a load is otherwise judged as wasteful, and would have been unlikely to happen in the case of a human performer. Inappropriate performance is not necessarily a problem, partly because practices, and thus what is considered appropriate, change all the time; but it does place automation in a particular position regarding the ways in which practices change. Accepting reduced garment endurance in line with the incapability of washing machines to treat garments separately could be an example. A range of more illustrative examples can be found in social gerontechnology research, where monitoring devices designed to take over part of the work of professional carers trigger alarms on the basis of misjudging situations, after which some users start to change their behaviours in order to prevent false alarms (Neven 2015; Hyysalo 2006). The nuance that co-performance adds to these observations is the specification of the particular type of delegations that pose a risk of leading to inappropriate artificial performance, namely, the delegation of polimorphic actions to artefacts. There is no routine distinction between polimorphic and mimeomorphic actions in the design practices that I am familiar with, but this analysis indicates that introducing these concepts may be fruitful for identifying more and less socially risky forms of automation.

Discussion and Conclusion

In this chapter, I have explored some of the implications of considering automated artefacts as co-performers of practices alongside people. I've drawn on the writings of Schatzki (2002, 2010), Reckwitz (2002a, b) and Shove et al. (2012) to argue that, with some stretching of terminology and ideas, it is possible to consider artefacts as performers of practices. To this I've added that in agreement with authors such as Pickering (1993), such artificial performance is only meaningful when part of social practices, and therefore always a case of co-performance alongside people. Subsequently, by analysing laundering as situated co-performance, and as a practice in which roles have been redistributed between humans and automated artefacts over time, I have explored what it means to extend the concept of practice performance to include artificial doings, including those outside of the direct presence and awareness of people, as part of it.

A key contribution of this chapter is the way in which it works through the implications of considering the specific category of automated artefacts, which derive their capability to 'do and say' from their electrically powered, computationally equipped artificial body/minds, as co-performers of practices. This thought experiment has revealed the importance of studying how and which practical knowledge and know-how gets embodied into artificial performers in the design process, as well as the limitations that the materialisation of previously situated judgments place on the capability of artificial performances to remain appropriate in varying circumstances. It has also highlighted the relevance of differences in capabilities between human and artificial body/minds for how practices are shaped, as well as for how practices tend to change towards suitability for artificial performance in processes of automation. Moreover, linking the notion of careers to artificial performance suggests that this transformation process is catalysed by the particular capability of artificial body/minds to relatively radically (as compared to humans) absorb practice-specific knowledge and skills over different 'generations'. Finally, the idea of differences in capabilities points to the delegation of what Collins and Kusch (1998) refer to as polimorphic actions to automated artefacts as a particular case of concern, because anthropological understandings of enculturation imply that automated artefacts are not capable of appropriately performing these. What is lost exactly when these actions are nonetheless delegated is a topic for further research.

In relation to existing practice theory literature, the thought experiment has illustrated how, compared to conceptualisations of materials as elements (Shove et al. 2012) and arrangements (Schatzki 2010) or artificial doings as background processes (Morley 2016), co-performance can offer a refined understanding of the role of automated artefacts and their designers in shaping situated performance. It also offers a novel perspective on processes of automation in relation to how practices change. However, many questions regarding the validity and potential of co-performance as a concept remain open.

The first set of questions relates to the theoretical positioning of the concept. How co-performance is positioned in relation to earlier considerations of distributed and artificial agency in theories of practice (Rose et al. 2003) and beyond—such as Latour's discussions of delegation (Latour 1992) and actor networks (Latour 2009) and Akrichs' (1992) notion of script—has not been worked out in detail. Further research is also needed into notions of enculturation and how developments in artificial intelligence might be conceptualised in relation to it.

Then there is another, more empirical set of questions. In this chapter, co-performance was worked through with a simplified example on the laundering of clothes, but automation is of course not the only reason why laundering has changed. Neither can the design of washing machines be separated from other practices of technology development. Interweaving of the concept of co-performance with other approaches to practice dynamics is required to make more careful links to existing work. Moreover, questions arise whether and how co-performance might apply to other practices within and beyond the home? Can a redefinition of 'good' practice in terms of artificial capabilities be identified beyond laundering practices as part of processes of automation, and if so, what are societal implications of such processes? Can the notion of inappropriate artificial performances be empirically linked to the delegation of polimorphic actions to artificial performers? And if so, (how) do inappropriate performances and their effects differ between human and artificial performers? Washing machines, like the central heating systems discussed by Morley (2016) and robotic vacuum cleaners discussed by Salvia et al. (2015) and Strengers (see Chapter 11), form relatively straightforward examples of automation, but what has exactly been delegated to home entertainment systems, smart phones or microwaves within what practices? How do we study practices in which multiple automated artefacts are involved? Can apps such as Facebook or Strava be viewed as artificial body/minds when they don't really have a body of their own? And can the distinction between the design process and situated performance still be made in cases of continuous software updates, such as discussed by Morley (see Chapter 9)?

Putting these questions aside for a moment, I would like to close this discussion by briefly speculating on the implications that co-performance might have for practices of designing automated artefacts. Because analysing co-performance involves zooming in on the detailed 'doings' of artificial body/minds, decisions made during the design process are foregrounded. By linking these decisions, through the practice-theoretic framework, to broader dynamics of social change, designers are offered a way into better understanding the broader implications of their detailed design decisions. Within the example case of laundering practices, it would be exciting, for example, to apply the concept of co-performance to reflect on emerging forms of automation in the practice. The Panasonic Laundroid[8] and the FoldiMate,[9] for example, present a next step in automating laundry care, namely, by adding the folding of clothes to the task set of artificial performers. How this innovation might transform laundering and related practices of wardrobe management and dressing is an interesting question. Analysing in detail how these artefacts handle and fold clothes differently than humans currently do, and considering which of the actions delegated to it require social awareness, offers a starting point to speculate about the potential consequences of this innovation. This analysis in turn could inform the design of future versions of the technology that might steer the practice in more desirable directions.

In conclusion, co-performance can, with some disclaimers, be argued to be a valid modification to the practice-theoretic framework that lifts automated artefacts out of the passive category of 'materials'. This chapter offers a new perspective on the role of automated artefacts in situated performances of laundering, as well as on how laundering has changed through cumulative automation. Future research on co-performance requires a refinement of the positioning of the concept in relation to other conceptualisations of the role of technologies in social change, fur-

[8] For example, shown in: https://www.youtube.com/watch?v=Gews8sQE9LY, and announced in: http://news.panasonic.com/global/press/data/2015/10/en151007-5/en151007-5-1.pdf, accessed September 15, 2017.

[9] See https://foldimate.com/, accessed September 15, 2017.

ther empirical work on analysing practices and their dynamics from a co-performance perspective and working through the implications of insights derived from such analyses for technology design practices.

Acknowledgements Writing this chapter would not have been possible without the support and feedback of Elisa Giaccardi and the book's excellent editors, as well as many discussions with a range of valued colleagues in the DEMAND Centre, at the University of Sheffield and in my own department in Eindhoven.

References

Akrich, M. 1992. The De-Scription of Technical Objects. In *Shaping Technology/ Building Society: Studies in Sociotechnical Change*, ed. W. Bijker and J. Law, 205–224. Cambridge, MA: MIT Press.

Bourdieu, Pierre. 1977. *Outline of a Theory of Practice*. Cambridge: Cambridge University Press.

Collins, Harry M., and Martin Kusch. 1998. *The Shape of Actions: What Humans and Machines Can Do*. MIT Press.

Cowan, R.S. 1983. *More Work for Mother: The Ironies of Household Technology from the Open Hearth to the Microwave*. New York: Basic Books.

Giddens, Anthony. 1984. *The Constitution of Society: Outline of the Theory of Structuration*. Cambridge: Polity Press.

Hyysalo, Sampsa. 2006. Representations of Use and Practice-Bound Imaginaries in Automating the Safety of the Elderly. *Social Studies of Science* 36 (4): 599–626.

Ingram, Jack, Elizabeth Shove, and Matthew Watson. 2007. Products and Practices: Selected Concepts from Science and Technology Studies and from Social Theories of Consumption and Practice. *Design Issues* 23 (2): 3–16.

Kottak, Conrad Phillip. 2005. *Mirror for Humanity: A Concise Introduction to Cultural Anthropology*. McGraw-Hill Higher Education.

Kuijer, Lenneke, and Matt Watson. 2017. 'That's When We Started Using the Living Room': Lessons from a Local History of Domestic Heating in the United Kingdom. *Energy Research & Social Science* 28 (Suppl. C): 77–85. https://doi.org/10.1016/j.erss.2017.04.010.

Kuutti, Kari, and Liam J. Bannon. 2014. The Turn to Practice in HCI: Towards a Research Agenda. In *Proceedings of the 32nd Annual ACM Conference on Human Factors in Computing Systems*, Toronto, ON, Canada.

Latour, Bruno. 1992. Where are the Missing Masses? The Sociology of a Few Mundane Artifacts. In *Shaping Technology/Building Society: Studies in Sociotechnical Change*, ed. Wiebe E. Bijker and John Law, 225–258. Cambridge: The MIT Press.

———. 1993. *We Have Never Been Modern*. Cambridge, MA: Harvard University Press.

———. 2009. A Collective of Humans and Nonhumans: Following Daedalus's Labyrinth. In *Readings in the Philosophy of Technology*, ed. D. Kaplan, 2nd ed., 11. Lanham, MD: Rowman & Littlefield.

Morley, Janine. 2016. Technologies within and Beyond Practices. In *The Nexus of Practices: Connections, Constellations, Practitioners*, ed. A. Hiu, T. Schatzki, and E. Shove. Routledge.

Neven, Louis. 2015. By Any Means? Questioning the Link Between Gerontechnological Innovation and Older People's Wish to Live at Home. *Technological Forecasting and Social Change* 93: 32–43.

Pickering, Andrew. 1993. The Mangle of Practice: Agency and Emergence in the Sociology of Science. *American Journal of Sociology* 99 (3): 559–589.

Poole, Fitz John Porter. 1994. Socialization, Enculturation and the Development of Personal Identity. In *Companion Encyclopedia of Anthropology*, ed. T. Ingold, 831–890. New York: Routledge.

Reckwitz, Andreas. 2002a. The Status of the 'Material' in Theories of Culture: From 'Social Structure' to 'Artefacts'. *Journal for the Theory of Social Behaviour* 32 (2): 195–217.

———. 2002b. Toward a Theory of Social Practices: A Development in Culturalist Theorizing. *European Journal of Social Theory* 5 (2): 243–263. https://doi.org/10.1177/13684310222225432.

Rose, Jeremy, Mathew Jones, and Duane Truex. 2003. The Problem of Agency; How Humans Act, How Machines Act. In *Proceedings of the International Workshop on Action in Language, Organisations and Information Systems (ALOIS-2003)*. Linköping University.

Rose, Jeremy, and Duane Truex. 2000. Machine Agency as Perceived Autonomy: An Action Perspective. In *Organizational and Social Perspectives on Information Technology*, 371–388. Springer.

Salvia, G., T. Cooper, T. Fisher, L. Harmer, and C. Barr. 2015. What is Broken? Expected Lifetime, Perception of Brokenness and Attitude Towards Maintenance and Repair. PLATE, Nottingham Trent University, 17–19 June 2015.

Schatzki, Theodore. 1996. *Social Practices: A Wittgensteinian Approach to Human Activity and the Social*. Cambridge: Cambridge University Press.

———. 2002. *The Site of the Social: A Philosophical Account of the Constitution of Social Life and Change*. Penn State University Press.

———. 2010. Materiality and Social Life. *Nature and Culture* 5 (2): 123–149. https://doi.org/10.3167/nc.2010.050202.

Shove, Elizabeth. 2004. Efficiency and Consumption: Technology and Practice. *Energy & Environment* 15 (6): 1053–1065.

Shove, Elizabeth, Mika Pantzar, and Matt Watson. 2012. *The Dynamics of Social Practice: Everyday Life and How It Changes*. London: Sage.

Warde, Alan. 2005. Consumption and Theories of Practice. *Journal of Consumer Culture* 5 (2): 131–153.

Williams, T.R. 1975. Socialization and Communication in Primary Groups: On the Origin of the Socialization Process. In *International Congress of Anthropological and Ethnological Sciences 1973*, Chicago, IL, 233–249.

11

Robots and Roomba Riders: Non-human Performers in Theories of Social Practice

Yolande Strengers

Introduction

Service robots have long been fodder for science fiction stories and future predictions, where they are commonly depicted as 'taking over' complex human tasks in both tantalising and sinister ways. As increasing numbers of domestic service robots enter the home, the idea of "humanoid assistants capable of *performing* multiple tasks and engaging in fairly sophisticated communication and interaction with people" (Forlizzi and DiSalvo 2006, 258; emphasis added) warrants further theoretical attention. Current visions for these robotic devices emphasise the goal of partial or complete delegation of the performance of practices to non-humans. However, the small number of empirical studies examining the entrance of robots into homes has tended to promote their 'uses' and benefits to humans, rather than reconsidering their changing position in social life. Given the increasingly autonomous roles envisaged for robots in contemporary society, this chapter considers their conceptualisation in theories

Y. Strengers (✉)
Centre for Urban Research, RMIT University, Melbourne, VIC, Australia
e-mail: yolande.strengers@rmit.edu.au

C. Maller, Y. Strengers (eds.), *Social Practices and Dynamic Non-Humans*,
https://doi.org/10.1007/978-3-319-92189-1_11

of social practice and explores the contentious idea that dynamic non-humans like service robots can perform practices in their own right.

Theories of social practice and their understandings of materiality already extend far beyond 'use'. In the posthumanist theories of practice drawn on in this chapter, materiality is encapsulated in the elements (Shove et al. 2012), material arrangements (Schatzki 2010), or resources (Reckwitz 2002a) that comprise practices and contribute to their integration, reproduction, and dissolution. Most commonly, devices like robots can be considered one of the material 'elements' of readily identifiable practice entities.[1] Practice theories (and the empirical analyses that follow from them) have thus tended to emphasise things which 'participate in' practices or are 'handled' by human practitioners (Reckwitz 2002a, 208). Alternatively, for Schatzki (2010, 130), materiality is expressed in "arrangements" which are "a set of interconnected material entities" comprising "humans, artifacts, organisms, and things of nature" to which human co-existence is tied.

Importantly then, material elements and arrangements include both humans and non-humans, such as bodies, animals, and other parts of the natural world, as other authors of this collection explore (see Chapters 2, 3, 4, 5). However, in the theories of practice mentioned above, it is individual people who "carry" or "host" practices (Shove et al. 2012, 12) or a "bodily and mental agent" (Reckwitz 2002b, 250) who carries them out. Moreover, most social practice theorists agree that practices comprise "embodied, materially mediated arrays of *human activity* centrally organized around shared practical understanding" (Schatzki 2001, 2, emphasis added). Therefore, while practices can be thought of as "materially mediated" or comprised of "human and non-human activities", performance is a role reserved for people (Schatzki 2001, 10; 2).

Social practice theories already have a variety of conceptual resources that readily account for the dynamism of smart, automated, and intelligent devices, as Morley convincingly demonstrates (see Chapter 9).

[1] Elements comprise materials, meanings, and competences in Shove et al.'s (2012) conceptualisation, or "forms of bodily activities, forms of mental activities, "things" and their use, a background knowledge in the form of understanding, know-how, states of emotion and motivational knowledge" for Reckwitz (2002b, 249).

Nonetheless, it is increasingly the case that humans are desiring and interpreting some devices—like robots and other forms of artificial intelligence (AI)—as capable of carrying out their own practices, complete with teleoaffective ends or goals (Schatzki 2002). Despite the fact that these 'ends' are designed by humans and embodied in human-generated algorithms, there does seem to be something interesting in exploring the idea that some non-humans can 'carry' and perform practices.

Such provocative suggestions raise a series of theoretical considerations that are important for our changing relationships with AI and other dynamic non-humans. For example, is the 'carrying out' or 'doing' of a practice enough to warrant special consideration of non-humans as performers or carriers of social practices, and all that entails? Is there anything unique about robots, as opposed to other 'materials' of practices, to justify this consideration? Where would we place humans in theories of social practice, if they are no longer the only possible performers? What might we gain—and lose—from interpreting social practices in this way? And what other non-human performers might we find along the way?

In what follows I consider such questions in relation to an increasingly popular service robot—autonomous floor cleaners, or 'robovacs'—drawing on available literature and marketing materials about the emergence of these devices into contemporary homes. Even though vacuum technologies have always autonomously done some of the 'doing' of vacuuming—by blowing, sucking, and extracting dirt and other materials from the floor—they have traditionally required a human performer, user, or operator. This person is assumed to make decisions about when and where to vacuum, how it should be done, and what frequencies of vacuuming are required, all of which reproduce (and are reproduced by) the social practice of vacuuming. In contrast, robovacs are designed to operate *without* a human operator, aside from initial purchase and setup (whether they do or not is still very much in question). In pursuing this line of enquiry, I uncover another type of dynamic non-human performing a distinctive practice, namely, domesticated animals that are apparently riding robotic vacuum cleaners for their own enjoyment.

The idea that robots or animals can perform practices is perhaps not as radical as it first sounds. Social practice theorists already accept that non-human 'material' performers are thoroughly enmeshed within the social world, consistent with "an ontology which treats materiality as part of society" (Schatzki 2010, 123). However, it is a bigger theoretical stretch to suggest that all non-humans have "bodily doings and sayings" (Schatzki 2002, 72)—a phrase that was coined with specific reference to human performers of practice. Nonetheless, some practice theories do lend themselves to the performative potentials I explore here, where devices (and, later in the chapter, animals) are 'doing and saying' something, even if that is embedded within the sociality of which they are a part.

Some may ask why bother going on this controversial journey at all, when there are other theories that readily account for non-human performativity and the "ontological diversity of actants" (Bennett 2005, 446)? Why not turn to a distributive theory of agency (Bennett 2005), actor-network theory (Latour 2005), or Barad's (2003) "posthumanist performativity", for example, the ideas of which have inspired and inform recent variations of theories of practice? Consider Latour (1987), who famously wrote about the delegation of the human door groom to an automated door to demonstrate how "programs of action" are entrusted and assigned from humans to non-humans. Such an account would no doubt prove fruitful in thinking through the apparent delegation of vacuuming from humans to robots. While these and other perspectives centre the role of materials and their performativity in accounts of human-material relations, the questions that concern me here relate specifically to the prevalent idea put forward by the robotics and AI industries that intelligent devices can perform practices autonomously without the need for a human performer. Despite—and perhaps *because* of—their emphasis on *human* performativity, the theories of practice I draw on in this chapter provide a specialised vocabulary and set of insights to conceptually explore the implications of this possibility. In turn, they pose questions back to theories of social practice regarding how to account for the intended (and actual) 'human-like' roles of non-humans. As empirical studies of social practices seek to understand the rapidly changing context of AI and other dynamic non-humans in everyday life, there is a need to carefully consider and account for these possibilities. Such conceptual teasing out may also be useful for those interested in steering or intervening in everyday practices involving emerging robotic devices and other forms of AI, such as those discussed below.

The chapter proceeds as follows. I continue by briefly outlining the growing diffusion of robovacs into households, situating these within broader technological utopian visions of technology delegation and service design. I then turn to three potential ways to conceptualise robovacs in theories of practice: first, as material elements of practices performed by people (as is consistent with some current theories of social practice); second, as (actual or perceived) autonomous performers of vacuuming practices; and third, as elements of *other* non-human practices performed by animals. In speculating on these ideas, my aim is not to prioritise one perspective over the other, but to consider how each draws our attention to different dynamics regarding the increasing diffusion of non-humans into everyday life. I conclude by considering the value of these conceptual roles for future research.

Vacuum Visions

The idea of a world in which machines do the work of humans has been a long-running feature of technological utopian imaginaries (Segal 1986). For example, early twentieth-century marketing for electricity emphasised how "our homes will be transformed to operate smoothly with the aid of electricity, so that the labour involved in cleaning, in heating, in cooking and in washing and other domestic tasks will be performed by an electrical deputy" (Whyte 1922 in Forty 1986, 190). Underpinning this common aspiration is the assumption that technologies are "neutral tools that function for us as detached, disinterested and unquestionably loyal servants" (Davison 2004, 86). In their study investigating representations of everyday life in future studies, Oksanen-Sarela and Pantzar (2001, 212) refer to this assumed-to-be desirable and possible assignment of performativity from humans to machines as a process of "cultural determinism", whereby it is "natural" to view technologies as replacing people in everyday life contexts. Aside from the many critiques of technological visions never quite achieving their desired outcomes, there is increasing recognition that many have, or are, coming to fruition, though not in the way their proponents may have intended (Dourish and Bell 2011). The emerging field of service robotics is one of these.

A decade ago Bill Gates (co-founder of Microsoft) outlined his vision for a "robot in every home" where "robotic devices will become a nearly ubiquitous part of our day-to-day lives" (Gates 2007, 62). The corresponding growth in robotic and automated devices provides some impetus to suggest that Gates' prophecy has, or is, being realised. One of the areas where there has been 'success' is the field of floor-cleaning robots, millions of which are now operating (or sitting unused) in households. Roughly 5.4 million service robots for personal and domestic use were sold in 2015, 16 per cent more than in 2014 (IFR 2016a). The International Federation of Robotics estimates that almost 30 million vacuuming or floor-cleaning robots will be sold between 2016 and 2019 (IFR 2016a).

Robovacs are "a logical merging of vacuum technology and intelligent technology" commonly featuring intelligent programming and sensors to autonomously brush, sweep, clean, and sometimes wash the floor (Forlizzi and DiSalvo 2006, 258). The sensors and intelligence embedded in these devices allow them to clean corners of a room, avoid falling down stairs or getting "stuck" in tricky places, focus attention on heavily used or "dirtier" areas of the floor, and "spontaneously adjust … [their] cleaning patterns in order to avoid running into moved furniture or crap on the ground" (Lufkin 2015).

As is immediately apparent from this list of things that autonomous vacuums can 'do', these devices are deeply indebted to past practices of vacuuming. By this I mean that they are informed by historic and current understandings of what makes vacuuming a practice or how it hangs together as a set of "doings and sayings" (Schatzki 2002). The robovac vision also closely resembles past utopian aspirations for technology and electricity, whereby our homes will be transformed to operate smoothly with the aid of electricity. For example, the International Organization for Standardization (ISO standard number 8373) defines a "service robot" as a device "that performs *useful* tasks for humans or equipment excluding industrial automation applications" (IFR 2016b). This 'usefulness' is defined in relation to humans, whom these devices are intended to serve. Delving further, the standard (ISO 8373) states that for service robots to be defined as such, they require "a degree of autonomy", which is the "ability to perform intended tasks based on current state and sensing, without human intervention" (IFR 2016b). This ranges from partial to full autonomy, the latter of which involves no active human-robot intervention.

The marketing of these devices reflects these ambitions. The iRobot Roomba promises its customers "effortless cleaning"; LG robot vacuums allows owners of its devices to "clean your floors without lifting a finger"; and a "roundup" review of different models claims they offer "near total set-and-forget convenience" (Brown 2014). These initial observations leave open the possibility of examining robotic vacuum cleaners from at least two mutually aligned perspectives: one which considers their 'usefulness' to humans and another which considers their ability to independently perform practices of service.

As far as service robots go, current research on robovacs indicates considerable room for improvement. In an early study of these devices, Forlizzi and DiSalvo (2006, 258) note that the iRobot Roomba Discovery Vacuum "pales in comparison to most robots, as well as to many other products already in the home", particularly because "it is not capable of planning or learning". Other studies emphasise the limited usefulness of these devices, the waning interest in them as the novelty wears off (Sung et al. 2009b), and, in some cases, their *dis*use or rejection from the household (Vaussard et al. 2014). Such reports suggest that, for the current state of affairs at least, robovacs are most similar to an appliance or device that readily fits within existing understandings of materiality in social practice theory, as I discuss next.

Robots as Material Elements

Most studies of robotic vacuum cleaners involving people and households have focused on how they are used or perceived (Fink et al. 2013; Forlizzi and DiSalvo 2006; Vaussard et al. 2014; Sung et al. 2009b) or how to increase householders' acceptance of service robots (Sung et al. 2009a). In doing so, they draw attention to the extensive human and bodily performances involved in robotic vacuuming (Sung et al. 2009c). Studies emphasise the importance of maintaining "user control" and designing devices that are "desired" by people and "better reflect user needs" (e.g. Sung et al. 2009a; Vaussard et al. 2014). This literature suggests that when robotic vacuum cleaners are studied "in the wild", they perform less like autonomous agents and more like "practical tools" (Vaussard et al. 2014) in a manner that is entirely consistent with common conceptualisations of non-humans in theories of social practice.

More specifically, these accounts reflect what Shove (2017, 159) calls the device-oriented role of some materials, which are "mobilised in practice" and "implicated in defining the practice itself". Conceptualising the robovac as a material element of vacuuming practices more than adequately explains the partial or complete disappearance of these devices from the practice as it becomes apparent that it is just another "cleaning tool with its own flaws" (Vaussard et al. 2014, 389).

When viewed as a material element of vacuuming, we can also understand variations of this practice being performed—either manually or automatically—as people switch between devices for different situations or circumstances. This is reflected in the considerable 'user involvement' that owners of robovacs report having to do. For example, Vaussard et al. (2014, 378) found that "most households had to make modifications to the physical layout of their homes in order to make the robot work well". More specifically, robotic vacuum cleaners required humans to perform regular maintenance such as removing hair or emptying waste from the capsule. Humans also assisted by moving things out of the way of the robot's path and occasionally manually moving the device to places and spaces it had missed cleaning. While this intensive interaction demonstrates considerable changes in the practice of vacuuming as a result of the robovac, there is clearly still a human performer at the helm of this practice arrangement—making sense of what to do with robots and integrating them within the existing constellation of elements required to perform vacuuming.

We can also understand the empirical findings from these and similar studies through Morley's (2017, Chapter 9) conceptualisation of automation within theories of practice. While Morley (2017, 81) acknowledges that practice theories have been limited when it comes to considering "what humans do collectively *beyond* their bodies", she conceptualises automated technologies as indirect material-practice relations or "extended processes" that shape the course and direction of social practices, even though they are not directly "handled" by human practitioners. In her analysis of automated factories and heating systems, Morley (2017, 83) focuses on "relationships that appear to matter *beyond* performance", thereby "zooming out" to consider automation part of a wider system of practices (Watson 2012), such as central heating or construction practices, in which human performativity remains clearly evident. Following this line of thinking, we could conceptualise robovacs as part of a system of housekeeping or domestic practices, wherein the actual 'doing' of vacuuming and other household practices is increasingly distributed to robotic devices.

From these conceptual starting points, we might continue to examine the entrance of service robots into everyday life and their reproduction, disruption, and transformation of practices. However, this is not the only way to examine and conceptualise the increasing diffusion and integration of robotic devices into everyday life.

Robotic Performers: Actual or Perceived?

As previously identified, the anticipated and desired goal for robotic devices is for them to perform practices independently of humans. In this section I explore the possibility that they can. A useful place to start is with the performative potential people themselves attribute to some non-human devices, which can be identified through the *relationships* they describe having with their robotic vacuum cleaners. For example, people give their devices names and characteristics, describing them as living beings such as pets or home helpers. In other words, they *imagine* them as human-like carriers of practices. For example, in one study of people's perceptions of and attitudes towards a robotic assistant in the home, research participants referred to them as "butlers", "servants", or "machines" (but stopped short of calling them "friends") (Dautenhahn et al. 2005). In another study (Forlizzi 2007), and on user blogs, robovacs are ascribed with gendered names and roles by their owners. For example, the following forum user comments on the vacuuming performed by their 'happy' male robot named Gort (Phoenix Rising 2014):

> *Gort happily cleaned my low pile carpet. He then moved onto happily cleaning my laminate flooring. While cleaning Gort managed to not eat up the curtains (a problem with the previous models) and he managed to not chew up the power cords (another problem with previous older models).*

Having an intimate relationship with a domestic service robot is also recommended by some scholars in order to "derive increased pleasure from cleaning", increase adoption and uptake, and help people manage unreliability (Sung et al. 2007, 145; 146). Similarly, Forlizzi and DiSalvo (2006, 261) found that people expect their domestic robots to be "intelligent", "gain knowledge of their environment over time, and adapt … [their] behaviour accordingly".

It is not uncommon for humans to anthropomorphise non-humans, but perhaps there is something more to learn from these 'relationships'. If a robot is viewed as a servant or butler, then humans are repositioned as owners or employers—a move which potentially redefines the *types* of practices humans are engaged in with these devices. For example, in a household with multiple service robots working in partnership with humans, we could think of people as performing household management practices (e.g. managing service robots that are 'doing' the cleaning), rather than carrying out discrete domestic practices such as vacuuming. From this starting point, we might investigate how non-humans and humans are co-performing a practice like robo-vacuuming (following Kuijer in Chapter 10) or performing two mutually dependent practices (i.e. household management and robo-vacuuming).

This notion of human-non-human relationships doesn't necessarily upset existing theories of practice and understandings of (human) performativity, if we accept that the social practice of vacuuming is changing in regard to the dynamics between humans and devices at the moment of 'integration' when performance takes place. Indeed it is *human* performers of vacuuming who are ascribing performative potentials to the robotic vacuum cleaner in this example. It doesn't necessarily follow that robots *are* performing practices, but it does seem useful to acknowledge that robovacs can take on a human-like role. Empirically investigating robots as performers or co-performers of practices could help make sense of these human desires and goals, the emerging service-like relationship between robots and people, and the role of these non-humans in redefining what vacuuming is and how it is carried out. Conceptually, we could arguably remain committed to the idea that only humans can perform practices, whilst still empirically studying and following what robovacs are 'doing' and 'saying', through human accounts and material observations.

Another way in which the performative possibilities of robovacs come to the fore is in situations where practices 'go wrong', and human bodies become disruptive material elements in the practice of robotic vacuuming—a role that is already well-accounted for in existing theories of practice (e.g. Shove et al. 2012). More specifically, human bodies can become physical materials that 'get in the way' of vacuuming, much like children's toys or furniture. This role is illustrated by the "horror story" of a South Korean woman whose hair got caught in a robotic vacuum cleaner when she fell asleep on the floor (Marszal 2015). The vacuum cleaner, carrying out its usual clean, encountered a bodily material (human hair) and sucked it up, causing considerable damage and distress to the sleeping woman. This case raised concern about the intersection between the practice of sleeping on the floor and robo-vacuuming, which are both common in South Korea. One way of conceptualising this intersection is as a practice 'bundle' of human-performed sleeping and robot-performed vacuuming which forms a "loose-knit pattern based on co-location and co-existence" (Shove et al. 2012, 17). The media reported this incident as a robot 'attack', in which the vacuum cleaner was held responsible for not recognising the sleeping body. Thus, observers of the situation blamed the robot, attributing it with performative potential, even though it is human designers and manufacturers who are ultimately responsibility for the robot's actions (and eventually get the blame).

This example again raises an important distinction between whether robovacs can *actually* perform their own practices and whether they are only *perceived* to perform them (by humans). Understanding human interpretations of robovacs as performers of practices may help to uncover the changing dynamics of vacuuming practices and the ways in which they transform household activities and relationships. For example, we could interpret the South Korean case as humans' anthropomorphising robovacs as performative through the competence that has been redistributed to these devices through design and automation (e.g. the 'doing' of vacuuming). This in turn redefines the meanings of robo-vacuuming as dangerous and sinister, rather than helpful and safe. In this sense, theories of social practice could more explicitly acknowledge where a non-human is perceived to be performing a practice, even if this is not what is actually going on.

Roomba Riding

In addition to thinking about service robots in the ways described above, there is a third and even more intriguing role these devices might take up in theories of social practice when considering their interaction with pets. Inspired by a recent phenomenon known as 'Roomba riding', we can conceptualise robovacs as being enrolled in practices of entertainment performed (and arguably invented) by domesticated animals.

Roomba riding refers to the pleasure some pets appear to derive from 'riding' a leading robotic vacuum cleaner known as the Roomba. As one article explains: "cats and dogs, for reasons unknown to mere humans, love hitching rides on the backs of robot vacuums" (Betters 2015). Of course, not all cats or dogs like riding (or can ride) Roombas, just as not all humans play football or like riding bicycles. Some actually appear scared of robovacs and stay as far away as possible (Forlizzi 2007). Nonetheless, a Google search on Roomba riding reveals hundreds of cat and dog videos and articles discussing this trend. In one reported case of Roomba riding, a woman walked in the door to find that one of her family's four chickens had "triggered the device and was in the middle of riding it through the living room" (Messenger 2015). Such accounts highlight how animals are inspired to ride Roombas for 'fun'.

What seems evident in these examples is a non-human performer—a domesticated cat or dog (or in at least one case, a chicken) developing a new entertainment practice involving robovacs. Stretching Schatzki's (2002, 72) claim that "what people, at almost any moment, are *in the first place* doing is whatever at that moment makes sense to them to do", we might contend that pets are also doing what "makes sense to them to do" in the presence of robovacs.

Of course, robots and animals are very different types of non-humans and call for different understandings of their roles in theories of practice that cannot be done justice here (see Chapter 4). Given that some animals are also social, and can pursue their own teleoaffective ends, it seems less controversial to suggest that they can be performers of practices compared to devices like robovacs. Nicholls, Maller, and I have considered the idea of non-human animal performers before in relation to the "curious consumers" (including pets and 'pests') that shape energy demand in the home (Strengers et al. 2016). In this work we conceptualised human-non-human interactions as an assemblage of practice, within which various non-humans and humans perform practices that intersect with each other. Such ideas may also be of value here. For example, in cases of Roomba riding, the robovac can be conceptualised as simultaneously enrolled in two interrelated practices—it is 'doing' the vacuuming whilst also being 'used' by another non-human as a "cat toy" (Darlin 2014). This seems consistent with the idea that objects have a "relational" role in practices (Rinkinen et al. 2015) and that these roles are "often ambiguous and always provisional" (Shove 2017, 164). Following Shove (2017, 156), the fluid material roles of the robotic vacuum cleaner acknowledges that devices can "fall into different categories ... depending on how they are positioned within and in relation to different practices".

We can also identify processes of practice circulation and recruitment (Shove and Pantzar 2007) facilitated by the robotic vacuum cleaner in this example, although in this case the performers being recruited are other non-humans, namely, domesticated animals. Similarly, the teleoaffective ends of Roomba riding are 'inspired' by non-humans—specifically vacuum cleaners and pets. We might even suggest that these practices colonise these non-humans' time and energy, in the same way that they can colonise peoples' (Shove et al. 2012).

However, it would be inaccurate to suggest that this assemblage is entirely composed of non-humans. Aside from the intimate human relations with robovacs described earlier in this chapter, we don't have to look far to find humans buying robovacs, filming their pets riding Roombas, dressing them in funny outfits (such as the famous shark costume video),[2] or posting videos of the experience on YouTube. Indeed, the term 'Roomba riding'—and the very definition of it as a 'practice'—has come from humans. New human entertainment practices are also generated or

[2] See https://www.youtube.com/watch?v=tLt5rBfNucc for video of cat in shark costume riding a Roomba robovac. As of 23 February 2018, the video had over 12 million views.

modified in this assemblage, such as watching "21 animals riding on Roombas to brighten your dreary day" (Diamond 2016). One argument, then, is that Roomba riding is still a social practice, carried, identified, and interpreted by people, but distributed and co-performed between robots, animals, and humans.

A more provocative interpretation would be to argue that new social practices can form *without* human intelligibility (but rather with animal and possibly even robotic intelligibility), a contention which challenges the distinct social ontology on which social practice theories are founded, namely, that "the social is a field of embodied, materially interwoven practices centrally organized around shared practical understandings" (Schatzki 2001, 3). For others, these ideas could provide tantalising opportunities to take a more flexible view of bodies, minds, and their intelligibility. For example, after Wheeler and Williams (2012, 5), we could redefine "mind" as "a property of systems (vegetative, animal, human) rather than of individual consciousnesses only".

Less controversially, examples such as this remind us that the home is a site inhabited by humans and non-humans which can perform multiple and fluid roles, and whose relationships can be understood in various ways. Other *living* non-humans clearly disrupt the flow and procession of practices, just as digital ones can and do. If we accept such propositions it becomes unhelpful to suggest that devices like robovacs are either materials of practices performed by humans, performers of vacuuming practices in their own right, *or* materials of practices performed by other living beings (e.g. pets). Instead, it is possible that they are all of these things at once or none of these things at all.

Discussion and Conclusions

In this chapter I have experimented with different conceptualisations of dynamic non-humans in theories of social practice drawing on the example of robovacs, their increasing diffusion into households, and their intended role to 'take over' the practice of vacuuming. First, I considered one of their 'traditional' positions as a material element in vacuuming practices; second, I explored the proposition that autonomous robots can be interpreted as performing their own practices; and third, I discussed the fluid and multiple roles that devices like robovacs can play in assemblages of practice such as Roomba riding, involving non-human practices performed by animals, people, and robots. Importantly, I have not sought to advocate one conceptual lens over another. Rather, I have undertaken this exercise to demonstrate how different theoretical perspectives foreground distinctive analytical questions and interpretations of what is going on. Each may prove interesting in different situations or at the same time, depending on what questions are being asked (and how these perspectives *inform* questions), and how the social practices of interest are "on the move" (Shove 2003).

A key contribution of this chapter has been to explore the variety of roles non-humans can occupy or take up in theories of social practice. As well as positing that "people in myriad situations adapt, improvise and experiment" (Warde 2005, 141), the perspective explored in this chapter invites consideration of the notion that non-humans can do the same (or something similar, or different). In some situations, I suggested that these roles are a matter of human interpretation, thus not challenging the notion that practices are performed by people. In the example of robovacs, this might involve recognising the human "routinized ways of understanding, knowing how and desiring" robovacs to be performers of vacuuming, by conceptualising and analysing them as such (Reckwitz 2002b, 250). However, in other situations I have suggested that some non-humans—like animals—could warrant consideration as performers of their own distinctive practices. In doing so I have left open the possibility that there may be some practices, like Roomba riding, that humans cannot 'invent' or directly reproduce, even if their existence as a recognisable entity is identified and conceptualised by people.

Importantly, it is essential not to lump all non-humans into the same bucket, as clearly there are not only different types of 'materials' in theories of practice but also different types of non-humans. Indeed, the perspectives explored in this chapter invite further consideration of the distinctions between different non-humans and their unique capabilities. The chapter also invites further reflection on how both non-humans and humans can be "defined, constituted and positioned with respect to each other through their role within specific practices" (Shove 2017, 157). As I stated in the introduction, it is perhaps unhelpful to look for concrete boundaries or definitions between non-humans that can or cannot perform practices. If we accept that *all* roles—both human and non-human—are provisional and relational, then this question is a matter for empirical enquiry and further conceptual experimentation.

On that note, the conceptual ideas explored in this chapter invite several different ways of investigating and intervening in the increasing prevalence of service robots and other dynamic non-humans like robovacs in everyday life. For example, if we are interested in what or how practices are changing around the robovac, we might begin by 'following' the practices performed by and around it, rather than investigating 'people performing robo-vacuuming'. By considering robots as performers of vacuuming, we are simultaneously invited to redefine what *people* are performing in relation to robots (e.g. household management or sleeping). This in turn potentially opens up a range of alternative questions and strategies to consider within this 'bundle' of human-robot practices. How, for example, might we make service robot management practices easier for human performers (if the aim is greater acceptance and uptake of robovacs)? Or how might we change practices of sleeping so that they don't interfere with robo-vacuuming? Conversely, how might robovacs distinguish between 'alive' and 'dead' non-human materials (the former of which they need to avoid, the latter of which they are designed to suck up)? While similar questions are arguably posed and explored in related theories (e.g. user experience), theories of social practice provide a distinctive ontology that could enhance understandings of human (and non-human) relationships and roles with robovacs and AI more broadly.

Likewise, by studying Roomba Riding in the home as a practice performed by non-human animals, we might better understand the corresponding set of *human* practices involved in entertaining and caring for pets, as well as be able to potentially anticipate and intervene in emerging digital consumption practices, such as new forms of animal play and entertainment involving devices originally intended for humans [such as iPad games for pets (Richardson et al. 2017)]. By only considering the *human* social practices associated with new and emerging AI, we effectively sidestep important questions about what animals might perform when they encounter them. This in turn misses opportunities to consider how technologies can become 'rebranded' as pet entertainment devices, in which both humans and animals are implicated. Importantly then, if robovacs or animals are too narrowly or broadly defined in theories of social practice, we potentially miss these other relational roles and dynamics.

While the analysis presented in this chapter is speculative, the increasing diffusion of service robots and other autonomous devices into homes warrants further attention of these ideas. However, they are likely to remain controversial terrain for a body of theory which privileges human performers or carriers. This does not mean we shouldn't embark on journeys into this territory. Indeed social practice theories provide a unique ontology that invite us to do just that. Such investigations are set to provide fascinating possibilities for understanding, negotiating, and intervening in the increasingly complex human-non-human relationships that characterise everyday life.

Acknowledgements This research was supported under the Australian Research Council's *Discovery Early Career Researchers Award* funding scheme (project number DE150100278). Thanks to Larissa Nicholls for her ongoing collaboration on this project. I am grateful to Janine Morley and Cecily Maller for providing insightful comments on previous drafts.

References

Barad, Karen. 2003. Posthumanist Performativity: Toward an Understanding of How Matter Comes to Matter. *Signs* 28 (3): 801–831.

Bennett, Jane. 2005. The Agency of Assemblages and the North American Blackout. *Public Culture* 17 (3): 445–465.

Betters, Elyse. 2015. Pets Riding Robot Vacuums. *Pocket-Lint*. Accessed October 9. http://www.pocket-lint.com/news/130167-pets-riding-robot-vacuums

Brown, Rich. 2014. The Robo-Race is on for the Housecleaning Crown. *Cnet*. Accessed October 9, 2017. https://www.cnet.com/news/robot-vacuum-roundup/

Darlin, Damon. 2014. Robot Vacuum Makes War on Cat Hair. *The New York Times*. Accessed October 10, 2017. https://www.nytimes.com/2014/01/23/technology/personaltech/review-the-roomba-880-from-irobot.html?_r=0

Dautenhahn, K., S. Woods, C. Kaouri, M.L. Walters, Kheng Lee Koay, and I. Werry. 2005. What is a Robot Companion—Friend, Assistant or Butler? In *2005 IEEE/RSJ International Conference on Intelligent Robots and Systems*, August 2–6, 2005.
Davison, Aidan. 2004. Reinhabiting Technology: Ends in Means and Practice of Place. *Technology in Society* 26: 85–97.
Diamond, Kelly. 2016. 21 Animals Riding on Roombas to Brighten Your Dreary Day. *Mashable Australia*. Accessed October 9, 2017. http://mashable.com/2016/04/07/animals-on-roombas/#hMKKf_PpEGqO
Dourish, P., and G. Bell. 2011. *Divining a Digital Future*. Cambridge, MA: The MIT Press.
Fink, Julia, Valérie Bauwens, Frédéric Kaplan, and Pierre Dillenbourg. 2013. Living with a Vacuum Cleaning Robot. *International Journal of Social Robotics* 5 (3): 389–408. https://doi.org/10.1007/s12369-013-0190-2.
Forlizzi, Jodi. 2007. How Robotic Products Become Social Products: An Ethnographic Study of Cleaning in the Home. In *Proceedings of the ACM/IEEE International Conference on Human-Robot Interaction*, Arlington, VA.
Forlizzi, Jodi, and Carl DiSalvo. 2006. Service Robots in the Domestic Environment: A Study of the Roomba Vacuum in the Home. In *Proceedings of the 1st ACM SIGCHI/SIGART Conference on Human-Robot Interaction*, Salt Lake City, UT.
Forty, Adrian. 1986. *Objects of Desire: Design and Society 1750–1980*. London: Thames and Hudson.
Gates, Bill. 2007. A Robot in Every Home. *Scientific American* 296: 58–65.
IFR. 2016a. Executive Summary World Robotics 2016 Service Robots. International Federation of Robotics (IFR).
———. 2016b. Introduction into Service Robots. International Federation of Robotics.
Latour, Bruno. 1987. Where are the Missing Masses? The Sociology of a Few Mundane Artifacts. In *The Social Construction of Technological Systems: New Directions in the Sociology and History of Technology*, ed. Wiebe E. Bijker, P. Thomas Hughes, and Trevor Pinch, 225–258. Cambridge, MA: The MIT Press.
———. 2005. *Reassembling the Social: An Introduction to Actor-Network-Theory*. New York: Oxford University Press.
Lufkin, Bryan. 2015. This Roomba Maps Your House So It Won't Keep Ramming into Your Couch. *Gizmodo*. Accessed October 9, 2017. https://www.gizmodo.com.au/2015/09/this-roomba-maps-your-house-so-it-wont-keep-ramming-into-your-couch/

Marszal, Andrew. 2015. Robot Vacuum Cleaner 'Attacks' South Korea Housewife's Hair. *The Telegraph*. Accessed October 10, 2017. http://www.telegraph.co.uk/news/worldnews/asia/southkorea/11399713/Robot-vacuum-cleaner-attacks-South-Korea-housewifes-hair.html

Messenger, Stephen. 2015. Woman Finds Chicken Riding Around Her House on a Roomba. *The dodo*. Accessed October 9, 2017. https://www.thedodo.com/chicken-rides-roomba-video-1472031925.html

Morley, Janine. 2017. Technologies within and Beyond Practices. In *The Nexus of Practices: Connections, Constellations and Practitioners*, ed. Allison Hui, Theodore Schatzki, and Elizabeth Shove, 81–97. London and New York: Routledge.

Oksanen-Sarela, Katja, and Mika Pantzar. 2001. Smart Life, Version 3.0: Representations of Everyday Life in Future Studies. In *Ordinary Consumption*, ed. Jukka Gronow and Alan Warde, 199–218. London: Routledge.

Phoenix Rising. 2014. Who has Tried Robot Vacuum Cleaners? Accessed October 9, 2017. http://forums.phoenixrising.me/index.php?threads/who-has-tried-robot-vacuum-cleaners.31700/

Reckwitz, Andreas. 2002a. The Status of the 'Material' in Theories of Culture. From 'Social Structure' to 'Artefacts'. *Journal for the Theory of Social Behaviour* 32 (2): 195–217.

———. 2002b. Toward a Theory of Social Practices: A Development in Culturalist Theorizing. *European Journal of Social Theory* 5 (2): 243–263.

Richardson, Ingrid, Larissa Hjorth, Yolande Strengers, and W. Balmford. 2017. Careful Surveillance at Play: Human-Animal Relations and Mobile Media in the Home. In *Refiguring Techniques in Digital Visual Research*, ed. E. Gómez Cruz, Shanti Shumartojo, and Sarah Pink, 105–116. London: Palgrave Macmillan.

Rinkinen, Jenny, Mikko Jalas, and Elizabeth Shove. 2015. Object Relations in Accounts of Everyday Life. *Sociology* 49 (5): 870–885. https://doi.org/10.1177/0038038515577910.

Schatzki, Theodore R. 2001. Introduction: Practice theory. In *The Practice Turn in Contemporary Theory*, ed. Theodore R. Schatzki, Karin Knorr Cetina, and Eike Von Savigny, 1–14. New York: Routledge.

———. 2002. *The Site of the Social: A Philosophical Account of the Constitution of Social Life and Change*. University Park, PA: The Pennsylvania State University Press.

Schatzki, Theodore. 2010. Materiality and Social Life. *Nature and Culture* 5 (2): 123–149. https://doi.org/10.3167/nc.2010.050202.

Segal, Howard. 1986. The Technological Utopians. In *Imagining Tomorrow: History, Technology and the American Future*, ed. Joseph J. Corn, 119–136. Cambridge, MA: The MIT Press.

Shove, Elizabeth. 2003. *Comfort, Cleanliness and Convenience: The Social Organisation of Normality*. Oxford: Berg Publishers.
———. 2017. Matters of Practice. In *The Nexus of Practices: Connections, Constellations and Practitioners*, ed. Allison Hui, Theodore Schatzki, and Elizabeth Shove, 155–168. London and New York: Routledge.
Shove, Elizabeth, and Mika Pantzar. 2007. Recruitment and Reproduction: The Carriers of Digital Photography and Floorball. *Human Affairs* 17: 154–167.
Shove, Elizabeth, Mika Pantzar, and Matt Watson. 2012. *The Dynamics of Social Practice: Everyday Life and How It Changes*. London: SAGE.
Strengers, Yolande, Larissa Nicholls, and Cecily Maller. 2016. Curious Energy Consumers: Humans and Nonhumans in Assemblages of Household Practice. *Journal of Consumer Culture* 16 (3): 761–780. https://doi.org/10.1177/1469540514536194.
Sung, JaYoung, Henrik I. Christensen, and Rebecca E. Grinter. 2009a. Sketching the Future: Assessing User Needs for Domestic Robots. In *RO-MAN 2009—The 18th IEEE International Symposium on Robot and Human Interactive Communication*, September 27, 2009–October 2, 2009.
———. 2009b. Robots in the Wild: Understanding Long-Term Use. In *Proceedings of the 4th ACM/IEEE International Conference on Human Robot Interaction*, La Jolla, CA.
Sung, JaYoung, Rebecca E. Grinter, and Henrik I. Christensen. 2009c. "Pimp My Roomba": Designing for Personalization. In *Proceedings of the SIGCHI Conference on Human Factors in Computing Systems*, Boston, MA.
Sung, Ja-Young, Lan Guo, Rebecca E. Grinter, and Henrik I. Christensen. 2007. "My Roomba is Rambo": Intimate Home Appliances. In *Proceedings of UbiComp 2007: Ubiquitous Computing: 9th International Conference, Innsbruck, Austria, September 16–19, 2007*, ed. John Krumm, Gregory D. Abowd, Aruna Seneviratne, and Thomas Strang, 145–162. Berlin, Heidelberg: Springer Berlin Heidelberg.
Vaussard, F., J. Fink, V. Bauwens, P. Rétornaz, D. Hamel, P. Dillenbourg, and F. Mondada. 2014. Lessons Learned from Robotic Vacuum Cleaners Entering the Home Ecosystem. *Robotics and Autonomous Systems* 62 (3): 376–391. https://doi.org/10.1016/j.robot.2013.09.014.
Warde, Alan. 2005. Consumption and Theories of Practice. *Journal of Consumer Culture* 5 (2): 131–153.
Watson, Matt. 2012. How Theories of Practice Can Inform Transition to a Decarbonised Transport System. *Journal of Transport Geography* 24: 488–496. https://doi.org/10.1016/j.jtrangeo.2012.04.002.
Wheeler, Wendy, and Linda Williams. 2012. The Animals Turn. *New Formations* 76: 5+.

12

Automation, Smart Homes and Symmetrical Anthropology: Non-humans as Performers of Practices?

Kirsten Gram-Hanssen

Introduction

This book raises the question of how to conceptualise dynamic non-humans within theories of practice, and whether they should be understood as performers, carriers or as other participants in the enactments of social practices. My specific interest in this question relates to concerns about energy consumption at a time when automation in homes is increasing. Within theories of practice, it has been established that to understand energy consumption we have to focus on the social practices which are carried by householders in everyday life (Gram-Hanssen 2011; Røpke 2009; Shove and Walker 2014). However, the idea of the fully automated home raises the question of whether this will continue to be true. Although a fully automated house is, so far, rather speculative, there are reasons to believe that we will see more of this, as I will describe below. If we can imagine a future in which houses demand and consume

K. Gram-Hanssen (✉)
Danish Building Research Institute, Aalborg University,
Copenhagen, Denmark
e-mail: kgh@sbi.aau.dk

C. Maller, Y. Strengers (eds.), *Social Practices and Dynamic Non-Humans*,
https://doi.org/10.1007/978-3-319-92189-1_12

energy without any humans being involved or even being present, then this raises the question of whether we should abandon conceptualisations of energy being consumed while accomplishing social practices, or whether we should develop theories of practice to also include the idea of automated devices or systems as carriers and performers of practices. My argument goes for the second, and the line of thought I will follow is inspired by the symmetrical anthropology of Latour (1993).

Latour's symmetrical anthropology does not include the performance of practices and neither does it differentiate simple technologies and devices from automated, intelligent or dynamic technologies and systems. In the Latourian actor-network perspective, all humans and non-humans have the same place and role in the network, and this network is all we have when seeking to understand the social world (Latour 1993, 1996). The role of the material within theories of practices, according to Schatzki (2010), is that material arrangements together with human practices form the nexus in which social phenomena exist. Material arrangements include all types of human and non-human entities, and these resemble what Latour describes as the actor-network. What I propose in this chapter is to follow the ideas of practice theory from Schatzki in understanding the social world as the interconnection of practices and material arrangements, *and* follow the ideas of symmetrical anthropology from Latour, assigning the same role for humans and non-humans, and thus including non-humans both as part of the material arrangements and as performers and carriers of practices: just as Schatzki includes humans as both part of the material arrangements and as carriers and performers of practices.

In developing this line of thought, I take a closer look at what automation is, especially within the home, and how it has developed over time. This historical approach shows a general co-construction of the social and the material within the everyday life of households. It also attempts to differentiate between types of technologies as regards levels of automation. This differentiation is useful for the discussion of relations between technologies as part of the material arrangements and technologies as dynamic non-humans performing practices. First, however, I develop the argument that home automation might be expected to expand in the future and thus why new ways of understanding this automation are relevant.

The Future Reach of Smart Homes and Automation

Visions of the fully automated smart home have existed for decades, although in real homes very little is happening. It seems there has been quite some economic interest from companies in developing these technologies, but that sales have so far been limited (The Economist 2016). History, however, presents several examples of how households were not always very keen on new appliances, such as washing machines and freezers, and that the adoption of these was slow in the beginning and only expanded after strong public and commercial pressure as well as extensive changes in material, social and cultural structures (Gram-Hanssen 2008). There are at present good reasons to believe that changes in energy provisioning systems in response to public response to climate change and other societal challenges will mean that public and commercial interests will join forces. Therefore, despite the relatively slow uptake so far, there are a number of other reasons to suggest that smart and fully automated homes may be more widespread in the future.

For decades, energy systems have been designed to produce according to demand, which typically means producing more in the evening when people return from work. With more renewable energy such as sun and wind power in energy systems, production systems are producing more intermittent generation. Moreover, as energy storage is expensive and wasteful, it is becoming relevant to ask whether demand can be adjusted to production. This technological development also depends on new digital possibilities of monitoring, and thus possibly controlling, electricity demand, either by households or by utilities (Wissner 2011). So far, flexibility in energy systems has mainly related to electricity. However, visions of energy systems relying 100 per cent on renewable energy argue for the integration of all types of energy, including gas and district heating. With these visions, buildings also become part of the flexibility in the system (Mathiesen et al. 2011). Furthermore, energy requirements in recent building regulations point towards increased "smartness" in houses. From only focusing on more airtight and better insulated buildings, today's tighter regulations call for more dynamic solutions for window shading, ventilation and different types of energy production and storage. These types of technology and the linkages between them may in themselves be reasons to have home automation to control the building.

If utilities are to persuade households to manage their energy, this will also require different smart home approaches, including digital metres with online feedback and possibilities to automate or control energy demand. It may also include persuading householders to invest in the digital technologies and infrastructures through other functions not directly related to energy management, such as entertainment, health and security: a phenomenon which has been called "funwashing" (Nyborg and Røpke 2013). Research on these approaches to smart homes suggests that the visions and programmes developed by utilities have quite a narrow understanding of householders and their everyday practices (Strengers 2013). There are not yet many actual evaluations of these smart homes. Where such evaluations exist, they suggest that smart home technologies are not easy to domesticate and that they influence the social roles in the households as well as the existing technologies in the home (Hargreaves et al. 2017). These evaluations also suggest that combinations of pricing schemes and smart technologies may influence a time-shift in energy consumption, although this also challenges the temporality of households' everyday practices (Friis and Haunstrup Christensen 2016). Finally they suggest that in general we have little knowledge of the real use of smart home technologies, and further development would benefit from user-oriented perspectives in technology development (Wilson et al. 2015).

Even though consumers appear unenthusiastic and social science research does not support technical optimism, there are good reasons to believe that smart homes will be part of the future. Therefore, there are also good reasons to develop understandings of how to reconceptualise these types of automation within theories of practices. However, before going further into a discussion of automation within theories of practice, a closer look at automation and smart home technologies is needed.

What Are Home Automation and Smart Home Technologies?

According to Oxford online learners' dictionaries "Automation" means "The use of machines to do work that was previously done by people",[1] where "Machines" are defined as "a piece of equipment with moving parts that is designed to do a particular job. The power used to work a machine may be electricity, steam, gas, etc. or human power".[2] Furthermore, Oxford online dictionaries define smart homes as "A home equipped with lighting, heating, and electronic devices that can be controlled remotely by smartphone or computer".[3] I will return to the adequacy of these definitions later, but for now they are a helpful start.

Looking at the types of domestic work previously done by people in their homes and taking a long historical approach includes what Schwartz-Cowan has called the shift from production to consumption, where tasks previously done by hand at home—such as milling flour and making clothes in the early industrialisation—were automated and taken out of the home and into industrial production (Schwartz-Cowan 1983). However, other types of work done in households stayed within the home, or moved from the market into the home, and were partly automated, including practices related to laundry (for an analysis of the automation of laundry, see Chapter 10). This partial automation of housework came together with the introduction of urban infrastructures of power, water and sewage systems. New appliances such as automatic washing machines which entered homes during the twentieth century thus have to be understood in relation to wider technological systems, including infrastructures (Schwartz-Cowan 1983). The massive integration of appliances in households in the post-war period also has to be seen together with wide-ranging changes in the social organisation of everyday life. New gender roles in homes were part of these changes, although

[1] https://www.oxfordlearnersdictionaries.com/definition/english/automation, accessed October 30, 2017.

[2] https://www.oxfordlearnersdictionaries.com/definition/english/machine_1#machine_1__4, accessed October 30, 2017.

[3] https://en.oxforddictionaries.com/definition/smart_home, accessed October 30, 2017.

there was no one-way causality between the growth in household appliances and women's place in the paid workforce (Schwartz-Cowan 1983). Rather it should be seen as two processes of change that were interlinked, in which women's enhanced entrance to the labour market also related to the possibility of keeping the family healthy, clean and comfortable with less time spent on work in the home.

The above definition of automation indirectly assumes that automation is doing the "same work" as previously done by humans, though what is happening with the introduction of new technologies and infrastructures in the home is most often that a previous task is substituted by something completely different. For example, keeping the house warm by carrying coal and lighting the stove was substituted by hot water coming through pipes and going into every room in the home to be heated. Keeping food refrigerated substituted different ways of preserving food that were integrated into the cooking and preparation of meals such as pickling and salting meat and vegetables. The practices of cooking and heating homes thus went through such profound changes that it may be relevant to ask whether these are indeed the same practices? Or should we consider them as practices dying out and others coming into existence?

Neither central heating nor refrigerators are automatic according to the definition of a "machine with moving parts". However, with the introduction of thermostats controlling the heat and cold, the central heating system or the refrigerator becomes automatic, and if later connected to information and communication technologies (ICTs), it may also become part of the smart home, as defined above, as well as the smart grid. Taking a historical approach, household technologies can be divided into different types relating to what type of human task they substitute. The first types of automation were about substituting human energy (or the use of animals) by combustion engines or wind/water mills. Later types were about substituting specific human activities such as washing clothes and later again about substituting human control of lighting and heating by thermostats, timers and sensors. This was followed by robots and the ability for machines to physically move around inside and outside the house. Finally, in the last step we also see the possibilities of automation substituting brain power and human communication. This is where the earlier smart home definition comes into use, stating that a smart home is equipped with devices that can be remotely controlled. In this definition, it is not explicit who or what performs this remote control; thus, it may be the home owner or other humans. It may also be controlled by a non-human system which builds on data gathering from metres and sensors that are controlled and used for data processing, allowing decision-making by humans to be substituted by self-learning algorithms and computer power (for further discussion on how to interpret software within theories of practices, also consult Morley, Chapter 9).

Whereas introduction of the electromotor in the last century made way for numerous new types of appliances such as vacuum cleaners and washing machines, ICT and the Internet of Things can be seen as making way for a new round of household automation and electrification (Røpke et al. 2010). Automation and smartness thus continue to build new types of automation on top of previous technologies and relate to substituting human energy, bodily movements, sensing, control, decision-making and communication, as shown in Table 12.1. Throughout history we have seen all types of housework being automated, including food preparation and cleaning, and the heating, cooling, ventilation and lighting used to maintain a comfortable indoor environment. Lately, we are also starting to see different types of surveillance to keep "an eye" on animals or (vulnerable) people in the household such as children and the elderly, to keep track of resource use or to monitor the home to prevent burglary.

This chapter raises the question of whether future home automation scenarios suggest the need to rethink the role of non-humans in theories of practices. It seeks to develop an approach which sees technologies as both part of the material arrangements *and* possibly as taking an active part in the performance of practices. To deal with this possible dual role, it is relevant to distinguish between different types of technologies to determine whether they are in the background of humanly performed practices or whether they are at the centre of performance itself. This question may be formulated as: Are there any paradigmatic shifts in the technologies introduced into the home and in what way do they relate to the practices performed? Before discussing this question further, the ideas of symmetrical anthropology and non-humans as carriers of practices will be developed to define when a human (or non-human) is carrying a practice and how this relates to the different types of automation identified in Table 12.1.

Table 12.1 Stages of household automation and substitution of human inputs

Substituting what human input?	Technical input	Examples of household automation
Human energy	Infrastructures of electricity, heat or gas	All appliances and automation rely on this
Bodily movements	*The previous plus* Electromotor	Washing machines, mixers, vacuum cleaners
Sensing	*The previous plus* Thermostats, timers, hydrostats, light and presence sensors, video recorders	Thermostatic control of ovens, control of specific indoor climate parameters, as either heat, ventilation or lighting, timers on all types of appliances, burglary alarms
Control and decision-making	*The previous plus* Computers and digitisation of sensors and other data	Robot vacuum cleaners and lawn mowers, integrated control of indoor climate based on different parameters
Communication	*The previous plus* Internet (of things), networked devices	Possible remote control of all connected appliances, including communication between different appliances, and machine learning in relation to this control

Note: This table shows how specific types of human input are substituted by different specific types of technical input and applied in different types of household automation, where all the later types of household automation always also include the older types of technical input

Symmetrical Anthropology and Theories of Practice

According to the ontology of Schatzki, the material arrangements together with human practices form the nexus in which social phenomena exist. In developing the role of non-humans in this ontology, Schatzki compares his ontology to that of ANT (actor-network theory) and Latour[4] (Schatzki 2010). Schatzki describes the difference between his ontology and that of Latour by saying that the network of human and non-humans which

[4] When Schatzki (2010) makes this comparison to ANT, he takes Latour as an example, though he does this in a broad approach, and rather than referring to any single texts from Latour, Schatzki writes in general about what Latour means. In my reference to the arguments from Schatzki, I thus do the same.

Latour describes resembles what Schatzki sees as the material arrangements, consisting of human bodies, technologies and natural elements (Schatzki 2010, 2011). The difference, according to Schatzki, is that for Latour this is all we have to include when seeking to understand the social world, whereas Schatzki argues that to understand the social we have to look at practices and how they are linked to these arrangements (Schatzki 2010). In another text, Schatzki further differentiates between his ontology and that of Latour and ANT by stating that his own ontology is within the position of social theory called contextualism, where ANT and Latour, according to Schatzki, are within nominalism (Schatzki 2002). The difference is explained:

> If, for example, a contextualist like myself opines that social orders exist only in particular contexts, a nominalist such as the actor-network theorist Callon (Latour, too) [*sic*] replies that all that exists are the orders themselves. (Schatzki 2002, 65)

My point of departure is to stay with the contextualism of practices, although I argue for enlarging it. Latour's most stunning contribution is the argument for a symmetrical anthropology, implying studying humans and non-humans by the same means and understandings (Latour 1993). Latour's argument for a symmetrical anthropology is built on a rejection of both the split between nature and culture and the split between our modern society and all the other cultures. Latour calls these the two Great Divides. When anthropologists go out in the wild to study other cultures, they study the relations between nature and culture and they see these as completely different from that of their own culture. In a symmetrical anthropology, we must do the same when studying our own world, and thus we cannot say that the natural part of our modern world is taken out of the equation and left for the hard sciences to describe as something completely separated from the culture.

> To become symmetrical, anthropology needs a complete overhaul and intellectual retooling so that it can get around both Divides at once by believing neither in the radical distinction between humans and nonhumans at home, nor in the total overlap of knowledge and society elsewhere. (Latour 1993, 101)

Schatzki is seemingly not convinced by these arguments. He argues against the language of symmetrical anthropology and opposes using the language of intentionality in relation to non-humans (Schatzki 2002, 199). Schatzki clearly states that in his ontology materiality is seen as an ingredient in social phenomena and that his ontology gainsays the separation between nature and culture (Schatzki 2010, 145), though at the same time he consistently talks of practices as only performed by humans (Schatzki 2010, 135).

In spite of this rejection of the symmetrical anthropology from Schatzki, this chapter is interested in how we can bring the thoughts of symmetrical anthropology into the ontology of theories of practices, implying that not only can material arrangements include both humans and non-humans, but that the performance of practices can equally be carried by both humans and non-humans. In Schatzki's understanding, a practice is sayings and doings held together by elements such as meanings, rules and ends (Schatzki 1996). Thus, it may be relevant to take a closer look at how Latour understands agency among humans and non-humans.

In his famous paper on "the missing masses", Latour (1992) gives a short but interesting introduction to what can be seen as the start of (smart) home technologies. In Latour's (1992) description, a door is a smart invention, which is needed when building houses and walls, otherwise one would have to break holes in the wall and rebuild every time one passes in and out the house. Doors, however, raise the problem of getting the door closed again every time someone has entered, otherwise the invention of the wall and the house may lose its relevance of shelter from a harsh outdoor climate and unwelcome strangers. This problem may be solved by educating people to close the door after passing through, by employing someone to keep the door closed or by delegating the task to non-humans, as seen with different types of automatic door closers. The problem is not, however, entirely solved in this way, as "when humans are displaced and deskilled, non-humans have to be upgraded and reskilled" (Latour 1992, 157). In following this line, Latour writes how this delegation from humans to non-humans includes non-humans delegating back prescriptions to humans. Prescriptions can be brought out in words, or we might say they can be brought out in rules that can be written, for instance, in textbooks or computer programmes. This delegation from humans to non-humans also includes a distribution of competences between humans and non-humans. With a hydraulic door closer, this distribution may incorporate having the physical strength to open the door, which may thus prevent children and the disabled entering the house, thus also implying a moral aspect of this delegation.

Within this Latourian perspective, energy consumption in a house (which strongly relates to the question of the door being closed after entering) implies understanding these relations and distributions between humans and non-humans and including the rules, moralities and competences of both humans and non-humans. This Latourian idea has also been drawn on in theories of social practice to look at how new tools for DIY work may imply new distributions of competences between the tool and the human (Watson and Shove 2008). Competences and rule following can thus be ascribed to non-humans as well as to humans. However, it is typically the question of meanings and ends where there is most disagreement between post-human theory and others (Schatzki et al. 2001). In the parliament of things (Latour 1993), non-humans such as cats, chimneys and ideas of comfort have a potentially equal say in the forming of the future smart home as have humans. In a Latourian perspective, the different types of technologies distinguished in Table 12.1 should thus not be treated differently from each other—they are all non-human actors and all of them have to be understood as having the same abilities as actors, independent of their technological sophistication. From a Latourian perspective, a smart home device controlling heating and windows, which can learn based on what residents and production systems do, has no more agency than a simple door or a hinge.

In my view, this is not quite satisfactory. When seeking to understand households' energy consumption and everyday life, a practice theoretical perspective has more to offer than an actor-network-theory perspective. The focus on the activities, rather than only the things and intermediaries, together with the inclusion of a teleo-affective element in holding practices (sayings and doings) together, gives more direction in an empirical investigation compared with ANT and its more limited focus on the linkages between all things. The practice theoretical perspective, by contrast, also conceptualises how practices are interlinked with each other and are routinised in temporal rhythms. In this way, the practice theoretical perspective better matches with an everyday life perspective in terms of what makes sense to people in their doings. In a more general way, Schatzki argues for the advantages of his practice theory compared to the ANT approaches by stating:

> Actor-network theory thereby fails to capture a key feature of human social life, namely, the practices that are tied to arrangements and help constitute social phenomena. Actor-network theory is also, as a result, unable to study relations between practices and material elements (…). In short, the bearing of materiality on human activity and social life lies not just in the constitutive and causal relations that hold between individual actors and particular objects, but also in how material entities are connected with temporally and spatially extended manifolds of organized human actions. (Schatzki 2010, 135)

However, with the practice theory as formulated here by Schatzki, we lose the symmetrical anthropology as formulated by Latour. Materiality and non-humans only have a place in the context of practices or as a result of practices and are not viewed as carriers of practices themselves. If non-humans were to be viewed as carriers of practices in a practice theoretical perspective, it could be relevant to distinguish between technologies as suggested in Table 12.1. This should involve a distinction between technologies that are only part of material arrangements and technologies with the dual role of possibly being both part of the material arrangements *and* performing practices. When discussing non-humans as carriers and performers of practices, it can be relevant first to distinguish between more and less complicated types of practices.

Non-human Performances in Dispersed and Integrated Practices

Schatzki has in some of his writings distinguished between dispersed and integrated practices (Schatzki 1996), where integrated practices are the most complicated in the sense that they consist of more elements holding the practice together. Examples of integrated practices in the home are cooking and washing practices, whereas examples of dispersed practices are asking, describing or, I would add, turning a switch on or off. Thus, dispersed practices can be elements in integrated practices and dispersed practices can be part of a multiplicity of different integrated practices. Furthermore, simple dispersed practices are not controlled by all the same elements as integrated practices and are typically not governed by teleo-affective structures (Schatzki 1996). Following this, it might be relevant to interpret some of the tasks performed by automated technologies in the home as dispersed practices performed by technologies. For instance, automation based on heat and light sensors giving input to simple if-then rule-based automation (*if* there are movements *and* it is dark *then* turn on light) could be seen as dispersed practices performing a simple task which takes part in many different integrated practices. This argument for non-humans as performers and carriers of dispersed practices might resemble what Kuijer (see Chapter 10) considers as co-performance by non-humans in practices, where the ends involved continue to remain those pursued by people. Thus, in her view, technologies can only be co-performers where humans are part of the performance as well and where humans are those pursuing ends.

In contrast to dispersed practices, integrated practices performed and carried by technologies would consist of different dispersed practices, linked together in forms which are guided by some form of teleo-affective structure or ends. If this teleo-affective structure is solely defined by humans, as Kuijer argues, then it would be wrong to interpret the integrated practice as performed by technologies. A core question then becomes to what extent technologies in themselves are developing ends? If smart homes can monitor and record their own performance within different material contexts (of technologies, human bodies, production system requirements and outdoor weather) and based on this can re-programme certain functionalities in the home, then I would argue that this might be relevant to consider as a type of practice performance, where the learning and re-programming may be considered as developing new ends. In following this line of thought, we can also start thinking about how, in performing these practices, the house can be seen to enrol humans living in the home in certain parts of the practice, as suggested by Strengers (see Chapter 11). However, these examples of smart homes are still rather speculative and only imaginable technological futures; thus, we need empirical studies to follow up on these ideas.

With a view to the technologies in Table 12.1, we might suggest that automation in its simplest forms should not be interpreted differently from other material arrangements, whereas some forms of rule-following sensor-based technologies could be interpreted as carriers of dispersed practices. This also implies that there are paradigmatic shifts in the technologies described in Table 12.1, where some of the latest types of automated technologies should be treated differently within a social science perspective. This is possible in a practice theory perspective, where these technologies not only form part of the material arrangement holding practices together, but also in themselves can be interpreted as performers and carriers of practices, developing new rules and ends as they perform the practices. The question of where the paradigmatic shift precisely should be in relation to technologies such as those depicted in Table 12.1 must be an empirical question.

Infrastructures as Performers of Practices and Consumers of Energy

In the previous sections I introduced the idea of simpler forms of automation as the non-human performance of dispersed practices, whereas integrated practices relate to automation such as self-learning algorisms. A smart home might thus be seen as carrying and performing practices within a wider material network including humans as well as non-humans. With some of the latest types of technologies introduced in Table 12.1, automation includes large interconnected systems of things and flows, which are controlled by automated intelligence and are possibly self-learning at a system level. Thus, questions of non-humans as carriers of integrated practices should probably also be understood at this infrastructural system level. Maybe energy should not only be seen as consumed by automated non-human practices performed in the home but also as energy conversions which are happening in a technological system including PVs, wind power plants, grids, controls, sensors, heat pumps and windows, as well as standards for what comfortable and adequate temperatures and healthy indoor environments are.

In a historical perspective, this could be seen in continuation of what Schwartz-Cowan called a shift from production to consumption, as pointed out previously. However, with continuous automation the actual doings which the householders are involved in may be still more limited and eventually disappear. If there are no doings and sayings by householders, can they then be interpreted as part of their practice? In the clothes-making industry, some practices in households disappeared as the making of the fabric and the sewing of the clothes were done in factories. However, households still have some doings left, particularly those related to being a consumer—the shopping act and wearing the clothes. In comparison, if the energy consumption related to heating the home is all controlled by the supply system, it may be difficult to say that there is still some carrying of the practice of consumption related to being warm. People may still do tasks related to being comfortable, for example, put on a sweater, but in a fully automated system of heating buildings, this may more closely resemble what is seen when we are outside, and the way humans adjust to natural variations in temperatures through their clothing.

In Schatzki's understanding the weather and the natural rhythm of seasons and circadian rhythm are viewed as processes within material arrangements, not as practices performed by non-humans. This is so even though it might be said that these processes also act upon and change the material arrangements and have a spatial and temporal element. Within these natural processes, there are many types of energy conversion, energy storage and energy consumption, although these are not what is normally considered within a practice theoretical perspective. In a time which has been called the Anthropocene, because of the wide-ranging impact which humans have on all "natural" processes, it is questionable whether we can make any sharp distinctions between "natural" processes and processes in technological systems which are designed and set up by human practices in interaction with material arrangements. And again, following the ideas of symmetrical anthropology, any sharp distinction between nature and culture is not ontologically adequate. Fully automated heating infrastructure systems may thus be understood from a practice theoretical perspective in similar ways to global climate and ecosystems, that is, as non-human performers of practices. Energy production will follow what is happening with the weather. When the wind is blowing and the sun is shining, renewable systems will produce energy and energy used to heat homes will be controlled against outdoor temperatures. Furthermore, the extent to which these systems will succeed in producing renewable energy will result in more or less impact on the global climate compared to a system also relying on the burning of non-renewable carbon-based materials.

From a practice theoretical perspective, it may be correctly argued that these possible smart systems and infrastructures are developed, constructed and maintained through the social practices of those working in the energy sector. However, is it helpful to only understand energy consumption in the context of the practices required to keep an energy system running? Elsewhere I have argued that including the practices of those working within the energy sector can be highly relevant when researching households' energy consumption (Gram-Hanssen et al. 2017). However, this is not the only relevant focus, even in a fully automated energy system. This is why the idea of seeing the systems as an infrastructural part of the material arrangements *and* as a performer of practices which is consuming energy is in itself a theoretical possibility to explore further.

Conclusion

In this discussion of the possible future of fully automated homes, I have suggested a possibility where households have very little to say and do in the control of the systems which support their home. My intention with this was not to imply that this is the only, the most desirable or the most realistic "smart" future. I have only assumed that it is a possibility and therefore relevant to discuss and theoretically conceptualise. I have elsewhere argued that the possibility of engaging with smart technology in the home may be a key part of the extent to which people will be interested in smart home technologies (Gram-Hanssen and Darby 2018). However, in many visions from the supply side, the assumption is that the less involvement from consumers, the better the possibilities of developing the system.

We do not know what will happen with the future of home automation. However, there are reasons to believe that we will have much more automation in our homes. Therefore, more evaluations of the smart technologies in use are relevant to inform and possibly guide the development. However, this also requires conceptualisations and theoretical approaches to understanding what automation means. In this chapter, I have taken some steps towards distinguishing between different types of automation, and I have raised the question of how to conceptualise automation within a practice theory perspective. I have used the Latourian argument of symmetrical anthropology to introduce the idea that non-humans can be carriers of practices as well as humans, just as both non-humans and humans can be part of material arrangements. The argument builds on attempts to historically list different types and levels of automation and link these levels of automation to different levels of simple and more complicated levels of practices. The arguments have pursued ideas of seeing some more simple forms of automation as non-humans performing dispersed practices, which can be interpreted as following rules but not including engagements and ends. These dispersed practices performed by non-humans can then take part in more complicated integrated practices performed either by humans or non-humans. At the level of integrated practices, the critical question for arguing that non-humans can be carriers and performers of practices is whether it is relevant to see not only rule following but also types of ends and engagements as properties belonging to the technologies. Here, the argument is that technologies which include intelligent systems with types of self-learning algorithms might be interpreted in this way. Finally, and especially related to questions of energy systems, the chapter has also discussed ideas of practices at a system level and has argued that this might resemble arguments that see global ecosystems as the non-human performance of practice.

Acknowledgements This chapter was written as part of the UserTEC project, financed by Innovation Fund Denmark. I am very grateful for comments and discussion during the work with good colleagues in the Research Group of Sustainable Cities and Housing, especially Mette Mechlenborg and Line V. Madsen, as well as very helpful comments from Yolande Strengers and Cecily Maller, as editors of this book.

References

Friis, Freja, and Toke Haunstrup Christensen. 2016. The Challenge of Time Shifting Energy Demand Practices: Insights from Denmark. *Energy Research & Social Science* 19 (Sept.): 124–133. https://doi.org/10.1016/j.erss.2016.05.017.

Gram-Hanssen, Kirsten. 2008. Consuming Technologies—Developing Routines. *Journal of Cleaner Production* 16 (11): 1181–1189. https://doi.org/10.1016/j.jclepro.2007.08.006.

———. 2011. Understanding Change and Continuity in Residential Energy Consumption. *Journal of Consumer Culture* 11 (1): 61–78. https://doi.org/10.1177/1469540510391725.

Gram-Hanssen, Kirsten, and Sarah J. Darby. 2018. 'Home Is Where the Smart Is'? Evaluating Smart Home Research and Approaches against the Concept of Home. *Energy Research & Social Science* 37 (Mar.): 94–101. https://doi.org/10.1016/j.erss.2017.09.037.

Gram-Hanssen, Kirsten, Nina Heidenstrøm, Gunnar Vittersø, Line Valdorff Madsen, and Mette Hove Jacobsen. 2017. Selling and Installing Heat Pumps: Influencing Household Practices. *Building Research & Information* 45 (4): 359–370. https://doi.org/10.1080/09613218.2016.1157420.

Hargreaves, Tom, Charlie Wilson, and Richard Hauxwell-Baldwin. 2017. Learning to Live in a Smart Home. *Building Research & Information* (Feb.): 1–13. https://doi.org/10.1080/09613218.2017.1286882.

Latour, B. 1992. Where are the Missing Masses? The Sociology of a Few Mundane Artifacts. In *Shaping Technology/Building Society: Studies in Sociotechnical Change*, ed. Wiebe E. Bijker and John Law, 225–258. Cambridge, MA: MIT Press.

———. 1993. *We Have Never Been Modern*. Harvard University Press.

———. 1996. On Actor-Network Theory: A Few Clarifications. *Sociale Welt* 47 (4): 369–381.

Mathiesen, B.V., H. Lund, and K. Karlsson. 2011. 100% Renewable Energy Systems, Climate Mitigation and Economic Growth. *Applied Energy* 88 (2): 488–501. https://doi.org/10.1016/j.apenergy.2010.03.001.

Nyborg, Sophie, and Inge Røpke. 2013. Constructing Users in the Smart Grid—Insights from the Danish eFlex Project. *Energy Efficiency* 6 (4): 655–670. https://doi.org/10.1007/s12053-013-9210-1.

Røpke, Inge. 2009. Theories of Practice—New Inspiration for Ecological Economic Studies on Consumption. *Ecological Economics* 68 (10): 2490–2497. https://doi.org/10.1016/j.ecolecon.2009.05.015.

Røpke, Inge, Toke Haunstrup Christensen, and Jesper Ole Jensen. 2010. Information and Communication Technologies—A New Round of Household Electrification. *Energy Policy* 38 (4): 1764–1773. https://doi.org/10.1016/j.enpol.2009.11.052.

Schatzki, Theodore. 1996. *Social Practices: A Wittgensteinian Approach to Human Activity and the Social*. Cambridge: Cambridge University Press.

———. 2002. *The Site of the Social: A Philosophical Account of the Constitution of Social Life and Change*. Pennsylvania State University Press.

———. 2010. Materiality and Social Life. *Nature + Culture* (Summer): 123–149 http://search.proquest.com/docview/619736105?OpenUrlRefId=info:xri/sid:primo&accountid=13552.

———. 2011. Where the Action Is (On Large Social Phenomena Such as Sociotechnical Regimes). *Working Paper 1*, Sustainable Practices Research Group.

Schatzki, Theodore, K. Knorr Cetina, and E. Savigny. 2001. *The Practice Turn in Contemporary Theory*. London and New York: Routledge.

Schwartz-Cowan, Ruth. 1983. *More Work for Mother: The Ironies of Household Technology from the Open Hearth to the Microwave*. Basic Books.

Shove, Elizabeth, and Gordon Walker. 2014. What is Energy for? Social Practice and Energy Demand. *Theory, Culture & Society* 31 (5): 41–58. https://doi.org/10.1177/0263276414536746.

Strengers, Yolande. 2013. *Smart Energy Technologies in Everyday Life—Smart Utopia?* Palgrave Macmillan.

The Economist. 2016. Where the Smart Is. *The Economist*.

Watson, Matthew, and Elizabeth Shove. 2008. Product, Competence, Project and Practice: DIY and the Dynamics of Craft Consumption. *Journal of Consumer Culture* 8 (1): 69–89. https://doi.org/10.1177/1469540507085726.

Wilson, Charlie, Tom Hargreaves, and Richard Hauxwell-Baldwin. 2015. Smart Homes and Their Users: A Systematic Analysis and Key Challenges. *Personal and Ubiquitous Computing* 19 (2): 463–476. https://doi.org/10.1007/s00779-014-0813-0.

Wissner, Matthias. 2011. The Smart Grid—A Saucerful of Secrets? *Applied Energy* 88 (7): 2509–2518. https://doi.org/10.1016/j.apenergy.2011.01.042.

Index[1]

A

[1] Note: Page numbers followed by 'n' refer to notes.

C. Maller, Y. Strengers (eds.), *Social Practices and Dynamic Non-Humans*,
https://doi.org/10.1007/978-3-319-92189-1

L

M

The manufacturer's authorised representative in the EU is Springer Nature Customer Service Centre GmbH, Europaplatz 3, 69115 Heidelberg, Germany. If you have any concerns regarding our products, please contact ProductSafety@springernature.com

Printed and bound by CPI Group (UK) Ltd, Croydon, CR0 4YY

30/09/2026

02225653-0009